INCENTIVIZING CHANGE

HOW GOVERNANCE REFORMS ARE CHANGING
THE URBAN LANDSCAPE OF BANGLADESH

Laxmi Sharma, Amit Datta Roy, and Melissa Alipalo

DECEMBER 2022

ASIAN DEVELOPMENT BANK

 Creative Commons Attribution 3.0 IGO license (CC BY 3.0 IGO)

© 2022 Asian Development Bank
6 ADB Avenue, Mandaluyong City, 1550 Metro Manila, Philippines
Tel +63 2 8632 4444; Fax +63 2 8636 2444
www.adb.org

Some rights reserved. Published in 2022.

ISBN 978-92-9269-984-0 (print); 978-92-9269-985-7 electronic); 978-92-9269-986-4 (ebook)
Publication Stock No. TCS220597-2
DOI: http://dx.doi.org/10.22617/TCS220597-2

The views expressed in this publication are those of the authors and do not necessarily reflect the views and policies of the Asian Development Bank (ADB) or its Board of Governors or the governments they represent.

ADB does not guarantee the accuracy of the data included in this publication and accepts no responsibility for any consequence of their use. The mention of specific companies or products of manufacturers does not imply that they are endorsed or recommended by ADB in preference to others of a similar nature that are not mentioned.

By making any designation of or reference to a particular territory or geographic area, or by using the term "country" in this document, ADB does not intend to make any judgments as to the legal or other status of any territory or area.

This work is available under the Creative Commons Attribution 3.0 IGO license (CC BY 3.0 IGO) https://creativecommons.org/licenses/by/3.0/igo/. By using the content of this publication, you agree to be bound by the terms of this license. For attribution, translations, adaptations, and permissions, please read the provisions and terms of use at https://www.adb.org/terms-use#openaccess.

This CC license does not apply to non-ADB copyright materials in this publication. If the material is attributed to another source, please contact the copyright owner or publisher of that source for permission to reproduce it. ADB cannot be held liable for any claims that arise as a result of your use of the material.

Please contact pubsmarketing@adb.org if you have questions or comments with respect to content, or if you wish to obtain copyright permission for your intended use that does not fall within these terms, or for permission to use the ADB logo.

Corrigenda to ADB publications may be found at http://www.adb.org/publications/corrigenda.

Notes:
In this publication, "$" refers to United States dollars and "Tk" refers to Bangladeshi taka.
All photographs, unless otherwise credited, were taken by Mohammad Rakibul Hasan.

Cover photo. Sharmin Sulatana, a councillor of Ghorashal *Pourashava*, holds a monthly awareness meeting on the dangers of child marriage and advocates for sending girls to school instead.

 Printed on recycled paper

Contents

Tables, Figures, Boxes, and Impact Stories — v

Foreword — viii
From the Minister of Local Government, Rural Development and Co-operatives, Government of Bangladesh — viii
From the Director General, South Asia Department, Asian Development Bank — x

Acknowledgments — xii

Abbreviations — xiii

Executive Summary — xiv

Summary of Key Messages — xxii

Project at a Glance — xxiv

Project Map — xxv

Introduction — 1

Chapter 1: Why Invest in *Pourashavas* — 7
Pourashava Challenges — 9
A Chance for Change — 16

Chapter 2: The UGIIP Strategy — 21
Project Preparation — 22
Project Design — 26

Chapter 3: The Latest Results — 37
Governance Improvements — 41
Infrastructure Improvements — 50

Chapter 4: Successes, Challenges, Lessons — 67

Chapter 5: Unfinished Business — 81

Appendixes 85
1. Key Documents 86
2. Key Interviews 87
3. The Evolution of Urban Governance Reform Indicators 89
4. Urban Governance Reform Indicators for Window A *Pourashavas* 95
5. Urban Governance Reform Indicators for Window B *Pourashavas* 103
6. Benapole, a Border Town, Grows with UGIIP Support 106
7. Estimated Investment Required Versus Project Allocation 110
8. Budget Reporting on Standing Committee on Women and Children 113

Mayors on UGIIP 119

Tables, Figures, Boxes, and Impact Stories

Tables

1	Distribution of 506 Urban Centers in Bangladesh	8
2	Local Government by Type	9
3	Project Investment Areas	38
4	Project Contributions to the Sustainable Development Goals	39
5	Payment Capacity of UGIIP-3 *Pourashavas*	42
6	UGIIP-3 *Pourashava* Spending on Basic Services and Public Satisfaction Rates	43
7	Volume of Training	44
8	Summary of Planned and Actual Improved Infrastructure	50
9	Summary of Improved Water Supply Infrastructure	58
10	Slum Improvements	64
A3	Comparison of the Urban Governance Improvement Action Plan between UGIIP-1, UGIIP-2, and UGIIP-3	89
A4	Urban Governance Improvement Action Plan Criteria for Window A *Pourashavas* under UGIIP-3	95
A5	Urban Governance Indicator Action Plan for Window B *Pourashavas*	103
A7	*Pourashava* Development Plan Budgets, Total Versus Allocated	110
A8.1	Status of Formation and Budgetary Activities for the Standing Committee on Women and Children in 35 *Pourashavas*, January–December 2020	113
A8.2	Status of Formation and Activating of Standing Committee on Poverty Reduction and Slum Improvement in 35 *Pourashavas*, January–December 2020	115
A8.3	Status of Formation of Slum Improvement Committees to Implement Slum Improvement Activities of 35 *Pourashavas*, January–December 2020	116

Figures

1	Key Areas of Government Improvement	xvi
2	Investment Summary	xxiv
3	Bangladesh—Urban Governance and Infrastructure Improvement Project *Pourashavas*	xxv
4	Seven Areas of the Governance Improvement Action Plan under the Third Urban Governance and Infrastructure Improvement (Sector) Project	27
5	Residents' Awareness and Participation	46

Boxes

1	Defining Governance	16
2	The Role of Central Government in Local Government Affairs	23
3	Recommendations from the Municipal Association of Bangladesh	78

Impact Stories

1	She Learned to Write Her Name with Project Group, Now She Does Its Banking	5
2	Child Marriages Decline with Project-Inspired Community Advocacy	13
3	Housing Brings "Different Kind of Strength" to Person with Disabilities, Widows, Veterans	18
4	Abuse Survivor Turns Award-Winning Entrepreneur	25
5	From Stay-at-Home Mom to Salon Owner	31
6	Mother of Five Builds Successful Shoe Business with $60 Loan	35
7	Water Connection Ends Rushing, Queuing, Joint Pains	40
8	After Livelihood Training, "Why Not College?"	49
9	After 35 Years of Working Construction, She Earns Equal Pay	53
10	Business Has Boomed with Safe, Dry Passage from New Road, Drainage	55
11	Infrastructure, Livelihood Trainings Bring Together Diverse Residents	56
12	With Access to Safe, Private Toilets, Women Move On to Other Causes	60
13	Club Promotes Female Athletics, Entrepreneurship, Leadership	62
14	She Sleeps Without Worry About Floods Washing Away Everything	65
15	Once Timid to Touch a Computer Key, She Is Training Hundreds of Others	70
16	Her Life Was "Once Unimaginable," Says Widowed Mother of Four	79
17	Learning to Sew, Save, Build Dream Homes Raises "Revolutionary" Voices	84

Greater mobility. The Urban Governance and Urban Infrastructure Projects have funded much needed infrastructure to make *pourashavas* more livable places.

Foreword
From the Minister of Local Government, Rural Development and Co-operatives, Government of Bangladesh

I am pleased to know that the Asian Development Bank (ADB) is publishing a knowledge product on the changes and lessons from the ADB-financed Urban Governance and Infrastructure Improvement (Sector) Project (UGIIP), a performance-based infrastructure financing project. I hope this publication will be useful to policy makers, development partners, and others implementing programs for improving their services to people.

Since ADB's emergence in the nation, it has provided funding to the Government of Bangladesh for important socioeconomic and governance reform initiatives. The bank has also been recognized for its cooperation, providing ongoing financial and technical support for numerous projects under the Ministry of Local Government, Rural Development and Co-operatives.

ADB has already aligned the country's partnership strategy with the government's Five-Year Plan (2021–2025) and its road map for promoting prosperity and fostering inclusiveness. Among the projects, UGIIP is concerned with the synchronization of urban governance, infrastructure development, and capacity development of the municipalities.

This concept of performance-based fund allocation has fostered sustainable development in Bangladesh for almost 20 years. With a comprehensive development outlook, it has driven the visionary transformation of Bangladesh's cities, despite the challenges of our times.

I hope this publication will contribute to the sustainable urban development of Bangladesh, to fulfill the dream of our Father of the Nation, Sheikh Mujibur Rahman, under the leadership of Sheikh Hasina, the Honorable Prime Minister. We believe this collaboration will consolidate ADB's ties with development initiatives of the Ministry of Local Government, Rural Development and Co-operatives.

I want to thank ADB for being a long-standing development partner of Bangladesh. The Government of Bangladesh will continue to work in partnership with ADB to address many other aspects of sustainable national development.

I wish every success for this publication.

MOHAMMED TAZUL ISLAM
Minister of Local Government
Rural Development and Co-operatives
Government of Bangladesh

Kotalipara Market. UGIIP-supported municipal markets are becoming a revenue source for the *pourashavas*. Kotalipara *pourashava* receives fees from shops, offices, and community centers rented inside the market.

Foreword
From the Director General, South Asia Department,
Asian Development Bank

The Asian Development Bank (ADB) and the Government of Bangladesh are celebrating more than 20 years of partnership in developing the country's pourashavas—important secondary towns where 40% of the urban population lives.

Starting in 2002, the Urban Governance and Infrastructure Improvement (Sector) Project (UGIIP) has expanded to 96 of the country's 328 *pourashavas*. As a trio of projects advancing urban development, UGIIP has brought transformative changes through governance, infrastructure, and socioeconomic development. UGIIP is one of ADB's most robust and effective efforts in strengthening urban governance.

UGIIP is an example of the development power of local revenue generation and public participation. As part of the reform process, *pourashavas* have demonstrated greater local revenue capacity to augment national allocations. Local revenues are funding locally drawn plans for social and environmental improvements. The local budget commitments are highly reassuring to the sustainability of the social gains under UGIIP for women, children, and the poor. Standing committees comprising *pourashava* officials and public stakeholders preside over planning in towns and wards, as well as advancing strategic causes such as gender equity, poverty reduction, and slum improvement.

Bringing dynamic changes to individual lives, communities, and towns through livelihood training, UGIIP has created new civic organizations, better environments for low-income communities, and infrastructure improvements in vulnerable residential and vital commercial areas.

I applaud the government's clear commitment to its secondary town development, enacted through the Local Government (*Pourashava*) Act of 2009, which institutionalized many of UGIIP's reforms. *Pourashavas* are in a better position to improve their tax collection systems and generate the nontax revenues that are important to building local sustainability and direct investment in socioeconomic initiatives. This law also grants *pourashava* residents a role in governance.

I also applaud the mayors of UGIIP-participating *pourashavas* for their willingness to govern differently—to govern better. They adopted new administrative processes and made themselves more available to their constituents, who engaged in meaningful ways in *pourashava* business and, as a result, feel greater ownership of their communities and towns. Through participation, accountability, and transparency, UGIIP mayors are champions of community development and good governance.

National decision makers should value UGIIP as a strategy that encourages performance-based budget allocations, giving cities and towns the right support to implement reforms. ADB remains committed to supporting *pourashava* development through improved governance, infrastructure, and social and economic opportunities.

KENICHI YOKOYAMA
Director General
South Asia Department
Asian Development Bank

Campaigning for girl education. Bhairab was one of many *pourashavas* to use new revenues in advocating education for girls, and against child marriage.

Acknowledgments

This knowledge product was led by Laxmi Sharma, senior urban development specialist for the South Asia Urban and Water Division (SAUW) of the Asian Development Bank (ADB), and Amit Datta Roy, senior project officer (urban infrastructure) in the Bangladesh Resident Mission, with guidance and support from SAUW Director Norio Saito and South Asia Department Director General Kenichi Yokoyama.

Much of the background knowledge and insights into the origins, development, and experience in designing and implementing the three-part Urban Governance and Infrastructure Improvement (Sector) Project (UGIIP) was made possible by the generous interview with Hun Kim, UGIIP-1 mission leader and former director general of ADB's South Asia Department (now retired); Masayuki Tachiiri, UGIIP-2 mission leader and the director of the Strategy, Policy, and Business Process Division of ADB's Strategy, Policy, and Partnerships Department; Norio Saito, UGIIP-3 mission leader and director of ADB's SAUW; and Alexandra Vogl, UGIIP-3 additional financing mission leader and principal planning and policy specialist in the Strategy, Policy, and Business Process Division of ADB's Strategy, Policy, and Partnerships Department.

For their support and guidance, the editorial team acknowledges the essential support of Rafiqul Islam, former senior project officer (urban infrastructure) of the Bangladesh Resident Mission, ADB; Sheryl Yanez, project analyst for SAUW; and Donna Marie Melo, administrative assistant for SAUW.

ADB is especially grateful to the Local Government Engineering Department (LGED) and UGIIP staff and consultants, past and present, for giving generously their time for candid interviews, ad hoc queries, coordination, and manuscript reviews. ADB is especially grateful for the time and attention given by AKM Rezaul Islam, the project director for UGIIP-3, for his reviews of this publication and support for the editorial team visiting the project sites; and to the following for their generous interviews: Md. Shafiqul Islam Akand, former project director for UGIIP-2 and UGIIP-3 and former additional chief engineer of the LGED; and Nilufar Yesmin, Jr., gender development and poverty reduction specialist.

Preparation of this publication benefited from reviews and insightful feedback from Alexandra Vogl. From the Bangladesh Resident Mission, comments were provided by Soon Chan Hong, senior country specialist; Pushkar Srivastava, project management specialist; and SA Abdullah Al Mamun, senior project officer (urban infrastructure).

The editorial team was led by Melissa Howell Alipalo (researcher, writer, and editorial production director) and supported by Mohammad Rakibul Hasan (photographer); Fabeha Monir (national editorial support for the impact stories); Jason Beerman, copy editor; Joe Mark Ganaban, layout and composition artist; Monina Gamboa, proofreader; and Ma. Cecilia Abellar, page proof checker. The team is grateful for the guidance and support of ADB's Department of Communications.

All photographs, unless otherwise credited, were taken by Mohammad Rakibul Hasan. The team is especially indebted to the UGIIP beneficiaries, who candidly shared their struggles with poverty and their gratitude for the UGIIP-motivated, *pourashava*-sustained support for women's economic independence, local development, and poverty reduction.

Abbreviations

ADB	Asian Development Bank
BME	benefit monitoring and evaluation
COVID-19	coronavirus disease
GAP	gender action plan
km	kilometer
LGED	Local Government Engineering Department
O&M	operation and maintenance
PDP	*pourashava* development plan
PWD	person with disabilities
PRAP	poverty reduction action plan
SIC	slum improvement committee
TLCC	town-level coordinating committee
UGIAP	Urban Governance Improvement Action Program
UGIIP	Urban Governance and Infrastructure Improvement (Sector) Project
WLCC	ward-level coordinating committee

Currency Equivalents

(as of 1 January 2022)
Currency Unit – taka (Tk)

Tk1.00 = $0.01
$1.00 = Tk85.73

Executive Summary

In 2002, the Urban Governance and Infrastructure Improvement (Sector) Project (UGIIP) introduced incentive-driven, performance-based lending that depended squarely on mayors leading a governance reform movement in their *pourashavas* (local government municipalities of mostly secondary towns and cities). As participating *pourashavas* met predefined, best-practice standards of good governance, they qualified for infrastructure funds. With this publication, the Asian Development Bank (ADB) and the Government of Bangladesh share their strategy of governance-driven urban development, with 20 years of experience in implementing UGIIP's evolving adaptations and results to improve its efficacy.

Context: Why Invest in *Pourashavas*

The economic hardships pourashavas were experiencing at the onset of UGIIP-1—and are still experiencing—will probably be familiar to many secondary cities across Asia and the Pacific. The growth rates of Bangladesh's urban populations have been twice the national rates for decades, according to the country's 2011 population census. Bangladesh's urban populations are widely disbursed across more than 500 smaller cities and towns. Despite their lower population sizes, these communities are important. Collectively, *pourashavas* represent more than 40% of the country's total urban population. They act as a buffer—an alternative destination—between the rural areas and the already overwhelmed larger cities.

The *pourashavas*' poverty and lack of economic growth, with poor infrastructure and services, are all symptoms of inadequate governance—with low community participation, deficient management, and local fiscal weakness.

Community participation. Many *pourashavas* have not had the institutional arrangements (e.g., committees, forums) in place for communities to collectively express their needs, preferences, and ideas to elected representatives and local officials. Municipal leadership has worked from an inherited history that has not involved the public in discussions about policies under consideration, budget development, and service delivery. As a result, top-down decisions have not benefited poor communities as much. This gap between public service and public needs has meant that elected officials have not been held accountable for their decisions and results (or lack of results) during their tenure. As government operations have typically lacked transparency, public expectations have lowered. The poor governance of pourashavas is most evident in the lack of key urban infrastructure and services: roads, water supply, sanitation, drainage, and solid waste management.

Poor management. *Pourashavas* in general are managed by many untrained staff, and approved local posts are often left vacant, pending appointments from the central level, because *pourashavas* generally lack the legal authority to hire locally. Most *pourashavas* do not have a land use or infrastructure development plan. Only the larger ("class A") *pourashavas* have an urban planner in the approved organizational chart, but the position often remains long vacant. Limited human resources are linked to poor planning and the precarious financial situation of *pourashavas*. Under this weak governance structure, *pourashavas* struggle without the incentives, knowledge, or capacity of elected leaders and *pourashava* officials to generate their own financial resources for making investments, providing staffing, and meeting public needs.

Pourashavas have also not had the capacity to generate sufficient revenues. *Pourashavas,* therefore, depend on budgetary transfers from the Government of Bangladesh, which account for more than half of the total revenues for most *pourashavas*. The *pourashavas* have not had the systems in place for effective tax collection that would have helped generate revenues for essential infrastructure and services. From 2000 to 2002, when UGIIP-1 was being conceptualized and designed, only 29% of *pourashavas* collected more than half of their estimated holding tax—their most important revenue source.

Fiscal decentralization. Under this weak governance structure, *pourashavas* struggle without the incentives, knowledge, or capacity of elected leaders and pourashava officials to generate their own financial resources for making investments, providing staffing, and meeting public needs. In 2013, local government spending accounted for only 8.2% of total government spending, representing only 1.1% of gross domestic product. From 1974 to 2013, local government spending increased only 7% in real terms.

The UGIIP Strategy

UGIIP challenges the status quo approach to urban governance and infrastructure improvement projects. ADB's experience has shown that infrastructure cannot be sustained unless the underlying governance problems are addressed first; so UGIIP focused on such governance issues, and their associated development challenges.

In preparing UGIIP-1, two significant concerns emerged: (i) the concentration of powers in the *pourashava* mayor; and (ii) the weakness of elected commissioners, local public officials, and community groups. From these two general findings, the design team behind UGIIP-1 knew the project needed to increase delegation, participation, and accountability within *pourashavas*. Good local governance is not possible without all three of these interdependent forces working together.

Project design basics. UGIIP may be best understood as offering to participating *pourashavas* three general opportunities: (i) financial incentives to achieve meaningful reforms, (ii) technical assistance and capacity building to implement the reforms, and (iii) performance-based financing for infrastructure that communities value and will also invest in. The opportunity to earn infrastructure funds was the incentive mechanism for reforms and a strategy for shared growth.

To develop *pourashavas'* capacity for delegation, participation, and accountability, the original project design team developed the Urban Governance Improvement Action Program (UGIAP). The UGIAP is the centerpiece of the UGIIP strategy and has been refined by each of the three UGIIP projects. The UGIAP under the ongoing UGIIP-3 covers seven key areas of good governance with as many as 28 sub-activities. *Pourashavas* that have participated in more than one UGIIP project (some have participated in all three) are evaluated against more stringent indicators.

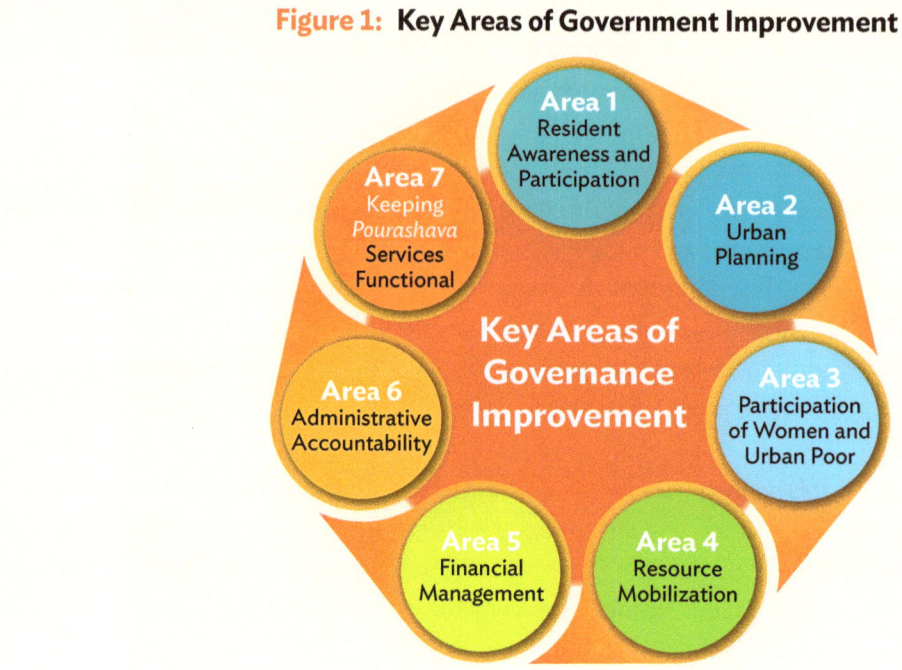

Figure 1: **Key Areas of Government Improvement**

Source: Urban Governance and Infrastructure Improvement (Sector) Project Management Office.

Infrastructure improvements. *Pourashavas* that meet the governance requirements receive their allocation for infrastructure improvements. *Pourashavas* could use UGIIP infrastructure financing on the following six types of infrastructure projects: (i) slum improvement, (ii) urban roads and drains, (iii) water supply, (iv) sanitation, (v) solid waste management, and (vi) municipal facilities. Municipalities that do not meet the requirements within the given time frame are disqualified from participating further in the project. Disqualification was a difficult lesson for only six *pourashavas* under UGIIP-1, and four *pourashavas* under UGIIP-2; however they rejoined later UGIIP projects and succeeded.

By linking reforms to financial disbursements for infrastructure improvements, the project motivates *pourashavas* to improve governance while also providing them with tangible infrastructure investments.

The Latest Results: UGIIP-3

To date, UGIIP has been—or is being—implemented in 96 of the country's 329 *pourashavas*. The results of UGIIP-1 and UGIIP-2 have been reported extensively in publications over the years; but as UGIIP-3 draws to a close, it is useful to assess its progress.

Governance Improvements

The major benefits expected from the reform actions were improved revenue collection, resulting in improved local services and local development. To build the capacity of project *pourashavas*, UGIIP-3 supported trainings on key local development topics and supported the shift to modern computerized accounting, billing, tax assessment, and communication (similar to UGIIP-1 and UGIIP-2). An estimated 38,000 people from 35 *pourashavas* participated in trainings.

Governance reforms fall into three broad categories of improved (i) financial capacity, (ii) service capacity, and (iii) residents' awareness of and participation in standing committees.

Improved financial capacity. UGIIP incentivized *pourashavas* to become more efficient in collecting property taxes and new local revenue resources—so they could pay their bills, repay loans, and reduce dependence on central government budget allocations.

UGIIP-3 *pourashavas* have exceeded the 85% target in holding tax collection efficiency. Since UGIIP-3 began, 35 *pourashavas* on average have nearly doubled the amount of holding and non-holding taxes collected; holding tax increased from $6 million to $13 million per year, and non-holding tax increased from $13 million to $22 million per year. Collection efficiency before UGIIP-3 ranged from 30% to 40%, but the average holding tax collection efficiency of all *pourashavas* reached 89.15% in fiscal year (FY) 2022 (ended June 2022). The additional revenues enabled the *pourashavas* to achieve the mandatory UGIIP reforms of paying overdue bills, remaining current on all electricity and telephone bills, and remaining on schedule with debt servicing; many accounts were overdue and in deep arrears.

The increase in local revenues has also created resources for funding local public services and development. UGIIP requires participating *pourashavas* to develop their own gender action plan (GAP), poverty reduction action plan (PRAP) (encompassing microcredit and skills training), and slum improvement plan. UGIIP goes further than typical urban development projects, by requiring *pourashavas* to identify funding sources for development plans in their annual budgets, and to show proof of annual disbursements during the project. The funding sources must be from local revenues and not central government transfers. This strategy aims to regularize *pourashava* allocations and spending, rather than creating a dependence on limited project-based funds or the unpredictability of central government budget transfers.

Improved service capacity. The success of these reforms, trainings, and technical assistance modernizing *pourashavas*' administrative systems, is evident in the residents' satisfaction. According to benefit monitoring and evaluation surveys, 80% of respondents had contacted their *pourashava* in the previous year. Most conducted business personally (63%); 34% conducted business through e-mail, letters, or other communication channels; and 4% by phone. Nearly 76% reported feeling either satisfied or highly satisfied with *pourashava* services and communication, and 88% said their issues or concerns had been resolved.

Public satisfaction is also evident in basic services. To modernize their operation and maintenance (O&M) of basic services, *pourashavas* spent UGIIP-3 project funds on a variety of equipment, such as garbage dump trucks, industrial street vacuums, and excavators. The mayor of Nilphamari *pourashava* and president of the Municipal Association of Bangladesh, Dewan Kamal Ahmed, reported that computer systems have limited corruption and helped to gain public confidence by reducing time, material, and financial losses.

Residents' awareness and participation in standing committees. UGIIP-1 established standing committees as mechanisms for involving the public more directly in municipal business, including planning the annual *pourashava* budget, managing community development plans, and implementing development activities. Project staff reported that these committees often ceased to function or did not function as robustly after UGIIP-1 and UGIIP-2 concluded. Under UGIIP-3, the Local Government (*Pourashava*) Act of 2009 legally requires *pourashavas* to establish the standing committees.

Resident charters and grievance redress. Each *pourashava* has established a resident charter that provides public information about *pourashava* services, costs of services, and municipal staff contact information. By June 2021, resident charters had been posted in the 36 UGIIP-3 *pourashavas*, bringing greater transparency and accessibility to *pourashava* services and staff. All 36 *pourashavas* have also established protocols to encourage the practice of residents expressing their opinions of *pourashava* services and holding officials accountable. Grievance redress cells have been formed in all 36 *pourashavas* and records are kept of complaints and resolutions. From July 2014 to June 2021, the project *pourashavas* received more than 15,000 complaints; some 12,000 complaints (75%) directly relating to their services were resolved.

Town-level coordinating committees and ward-level coordinating committees. All 36 *pourashavas* have established town-level coordinating committees (TLCCs) with 50 members in each TLCC and at least 34% women's participation. Ward-level coordinating committees have been established in all 343 wards and, interestingly, have an even higher female participation rate of at least 40%, with 15% of members coming from low-income communities. UGIIP staff have described the town and ward committees acting as a "mini parliament" in which residents and neighbors can bring ideas and concerns about *pourashava* activities and service delivery. During UGIIP-3, more than 70% of TLCC members attended most meetings, held at least once every 3 months.

Women and children's affairs committees. During UGIIP-3 implementation, the 36 *pourashavas* have allocated from their own revenue sources more than $3.7 million to support the implementation of GAPs. In total, around 12,000 low-income women from the 36 *pourashavas* have received training on different skills development, and around 37% of them have gained full-time employment.

Poverty reduction and slum improvement committees. *Pourashavas* have also allocated from their own revenue sources $8.4 million for poverty reduction. Slum improvement committee (SIC) members are predominantly women (comprising 86.13% of the total membership). These committees manage the implementation of slum improvement projects, covering things such as water supply, sanitation, footpaths, drainage, and solar lights. The project has trained almost 1,400 women and 46 men in the SICs participating in UGIIP-3.

Infrastructure Improvements

Through collaboration with standing committees, the UGIIP-3 *pourashavas* developed publicly agreed *pourashava* development plans (PDPs), based on prioritized lists of infrastructure projects. Following community-based development principles, beneficiaries were involved in the planning, training, and implementation of the selected committees' plans, activities, and projects. In most cases, initial estimated costs of the development plans exceeded the expected UGIIP-3 funding allocations, requiring *pourashavas* to develop revised development project proposals. In total, UGIIP-3 funded an average 42% of the initial PDP proposal costs, an indication of the need *pourashavas* have for infrastructure. Supported by ADB's

Urban Climate Change Resilience Trust Fund, subprojects incorporated climate-resilient materials and construction practices into their design. Roads and drainage are designed with 10% additional capacity for increased rainfall because of climate change.

The infrastructure projects were scheduled over three phases. The first phase concentrated mostly on roads and drainage. Most of the other infrastructure projects were built in the second and third phases. Under UGIIP-3, as of September 2022, infrastructure included:

(i) 1,727 kilometers (km) of roads improved or rehabilitated;
(ii) 632 km of drains built or improved for flood management;
(iii) 390 km of pipes installed or upgraded for water supply with 85,230 individual meters;
(iv) slum improvement subprojects completed in 262 slums;
(v) sludge management facilities either completed or in progress in 23 project towns;
(vi) solid waste disposal sites built, or being improved, in 26 project towns (handling 300 tons per day);
(vii) climate issues considered in designs and incorporated in all subprojects;
(viii) about 15,000 tons of carbon dioxide emissions reduced per year; and
(ix) municipal buildings and facilities such as bus terminals, parks, walkways, and boat-landing stations.

Qualifying infrastructure. *Pourashavas* could use UGIIP infrastructure financing on six types of projects: (i) slum improvement, (ii) urban roads and drains, (iii) water supply, (iv) sanitation, (v) solid waste management, and (vi) municipal facilities.

Successes, Challenges, Lessons

From operational experience, ADB's staff, consultants, and contractors can share knowledge of factors making these projects successful, including good practices—as well as the challenges and lessons learned from their design and implementation—to benefit those interested in making urban development more efficient, impactful, and sustainable.

The UGIIP approach has matured well, continuously improving over more than 20 years of implementation. All three UGIIPs have shown what works, what does not, and what still needs attention. The success factors should encourage project designers and governments to raise expectations of what local governments can accomplish. UGIIP has proven that limited resources and capacity can be overcome with support and public accountability.

There have been perennial challenges that seem to have been resolved only nominally and incrementally over the years, such as (i) issues with central government-centric policies on the management of local government human resources, (ii) a shallow national consultant pool, and (iii) the prioritizing of O&M. More ideas and innovations are still needed. Many of the UGIIP-3 *pourashavas* represented a new challenge because they were smaller and less developed. Most of the *pourashavas* that have not yet participated in UGIIP are lower-class municipalities and will require more capacity development and technical support.

The Successes, Challenges, Lessons chapter covers the following insights:

1. What mayors initially feared has been their greatest success: public accountability
2. Designing for results and momentum offers mayors a chance at good results within their political tenure
3. Well-structured evaluation of a *pourashava's* performance protects the credibility of the entire program
4. Performance-based access to funding can create healthy competition
5. Intensive project support locally can elevate ceremonial governance into effective governance
6. A visible municipal presence overseeing contractors is the best quality assurance
7. Small contractors need fair competition
8. Budget allocations are proof of *pourashava* intent and actual action
9. Mitigate disruptions to implementation schedules and regain momentum with strategic teamwork
10. Demand and dependence can be addressed by competition and cost sharing
11. Operation and maintenance requires special emphasis
12. *Pourashavas* and projects need more responsive human resource protocols and a robust consultant marketplace to recruit and retain talent
13. Town planners are needed
14. Introduction of volumetric meters and tariffs is a tough challenge
15. Slum communities need greater equity in development spending
16. Recognize and promote women's role and contribution to development
17. Standing committees are vulnerable to integrity gaps
18. Decentralize municipal staffing decisions

19. Leverage the law on standing committees
20. Encourage systematic knowledge sharing and transfer of good practice

The discussion includes recommendations from the Municipal Association of Bangladesh, an enthusiastic supporter of UGIIP.

Unfinished Business

With UGIIP-3 drawing to a close in 2022, ADB and the government are determining how best to serve the governance and infrastructure needs of people living in *pourashavas*, especially those that have not yet participated in UGIIP. One thing is certain: each UGIIP has expanded its coverage across the country, raising expectations of the government to continuously reform its governance practice.

Project directors and mayors hope future UGIIP or UGIIP-inspired programs will increase funding and include more *pourashavas*. The country has enough experience with incentive-based reforms and infrastructure funding to lead a national program supported by development partners.

UGIIP has demonstrated to the government how it could leverage its national budget transfers to incentivize reforms and reward *pourashavas* according to their performances. ADB could assist in leveling the competition field between *pourashavas* by supporting a national program with additional infrastructure financing and technical support. Certainly, a national program requires more development partner support to (i) replicate the UGIIP strategy in *pourashavas* that have not yet had the opportunity to participate and (ii) to continue supporting those *pourashavas* that have done well with UGIIP, but require further monitoring, capacity building, and infrastructure finance.

UGIIP-1 supported the government in developing a draft urban policy, though it has not been approved. Further advocacy for the policy in UGIIP-2 and UGIIP-3 has not achieved much progress. Without a national urban policy and the adoption of performance-based budget allocations to *pourashavas*, the country's important secondary towns may continue to rely on project-based opportunities for infrastructure improvements, capacity building, and social development support.

Summary of Key Messages

1. **Performance-based allocations.** In 2002, the Asian Development Bank introduced a new strategy of performance-based infrastructure funding to incentivize municipalities in Bangladesh to undertake governance reforms, which would position them with greater fiscal and administrative capacity and public accountability. As municipalities significantly improved their governance, they qualified to receive their allotted funds for infrastructure improvements. This performance-based, governance-driven strategy for infrastructure improvements has been improved over three projects of the Urban Governance and Infrastructure Improvement (Sector) Project (UGIIP) from 2002 to the present. It is an adaptable strategy that can be replicated in countries and cities to effect parallel progress in governance and infrastructure.
2. **Greater fiscal autonomy through local revenue generation.** The tendency for municipalities to over-rely on central government fiscal transfers can be relieved through more effective local taxation and nontax revenue streams, but municipalities need technical support in improving their tax administration and developing local revenue sources. By the end of UGIIP-2, in 2016, as UGIIP-3 was being designed, local revenue generation in project *pourashavas* had increased by 106% ($4 million) for holding taxes and 129% ($11.5 million) for non-holding taxes over their baselines. This is one of the most important of the UGIIP-introduced reforms for strengthening municipality self-reliance. Without local revenues, *pourashavas* have been at a development standstill or have even reversed course as infrastructure has continued to depreciate while *pourashava* populations have grown, leading to the proliferation of slum conditions in low-income communities.
3. **Governance requires accountability through shared decisions.** Residents' participation in urban governance improves elected leaders' and officials' responsiveness and sense of public accountability. The mechanism for participatory governance comes from establishing town and ward coordinating committees comprising officials and residents, who work together to identify local infrastructure priorities. Specialized committees take on various administrative and operational areas, such as gender equity, slum improvements, and poverty reduction. These committees develop action plans that are funded by the annual budget, in which the town-level standing committee has a direct role.
4. **Livelihood training for women is a multiplying factor.** A 1-month training program to develop a bankable skill and business sense can empower a woman to pursue entrepreneurship, reinvest in other women, and continue to grow and diversify her business pursuits. The opportunity to engage in such trainings under the guidance of a female trainer was a key factor in stories told by many female participants. Economically empowered, some women also became socially empowered to challenge practices and norms that undermine female prosperity and safety, such as child marriage, dowry traditions, domestic abuse, and unequal pay for equal work. Women organized themselves into microfinance cooperatives to protect a portion of their assets and provide capital for other women's business ventures.

"Urban Governance is the sum of the many ways individuals and institutions, public and private, plan and manage the common affairs of the city."

—UN-HABITAT

Project at a Glance

Executing Agencies: Local Government Engineering Department and Department of Public Health Engineering
Implementing Agencies: *Pourashavas*
Project Titles:

    Urban Governance and Infrastructure Improvement (Sector) Project
    Second Urban Governance and Infrastructure Improvement (Sector) Project
    Third Urban Governance and Infrastructure Improvement (Sector) Project

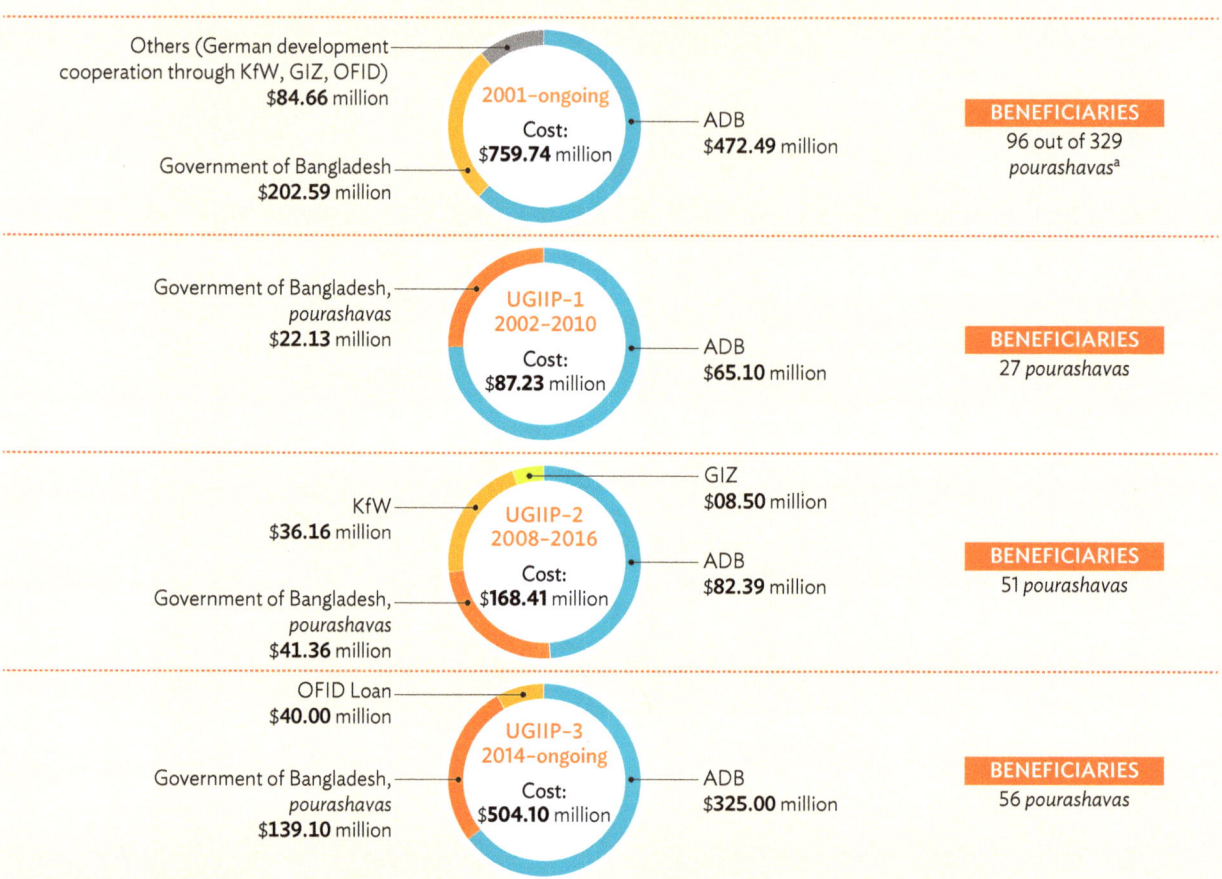

Figure 2: Investment Summary

2001–ongoing — Cost: $759.74 million
- Others (German development cooperation through KfW, GIZ, OFID) $84.66 million
- Government of Bangladesh $202.59 million
- ADB $472.49 million
- BENEFICIARIES: 96 out of 329 *pourashavas*[a]

UGIIP-1 2002–2010 — Cost: $87.23 million
- Government of Bangladesh, *pourashavas* $22.13 million
- ADB $65.10 million
- BENEFICIARIES: 27 *pourashavas*

UGIIP-2 2008–2016 — Cost: $168.41 million
- KfW $36.16 million
- Government of Bangladesh, *pourashavas* $41.36 million
- GIZ $08.50 million
- ADB $82.39 million
- BENEFICIARIES: 51 *pourashavas*

UGIIP-3 2014–ongoing — Cost: $504.10 million
- OFID Loan $40.00 million
- Government of Bangladesh, *pourashavas* $139.10 million
- ADB $325.00 million
- BENEFICIARIES: 56 *pourashavas*

BENEFITS

ECONOMIC COMPETITIVENESS
- Improved water supply: 419 km of piped supply, 60,000 individual meters
- Improved roads: 3,267 km
- Various municipal facilities (82 new facilities, e.g., parks, markets, community centers, bus terminals)

EQUITY AND INCLUSIVENESS
- Upgraded 262 slums
- Income-generating training, support for women and vulnerable people

ENVIRONMENTAL SUSTAINABILITY AND RESILIENCE
- Fecal sludge management in 23 sites
- Solid waste management at 33 disposal sites, with daily capacity of 200 tons
- Urban drainage: 724 km
- 15,000 tons of carbon emissions reduced annually after UGIIP-3 implementation

ENABLING
- Capacity development for urban service delivery, planning, and financial management
- Policy reforms implemented and legalized through the Local Government (*Pourashava*) Act of 2009
- Urban master plans and annual budget developed through local, pro-poor, gender-inclusive processes
- PDPs, GAPs, and PRAPs funded annually from own revenue sources
- O&M plans funded from own revenue sources
- Computerized tax records and functional billing systems

ENGAGEMENT
- Standing committees established—comprising *pourashava* officials, residents, and special representatives of vulnerable communities—to ensure public participation in setting annual budgets, and implementing development plans

ADB = Asian Development Bank, GIZ = Deutsche Gesellschaft für Internationale Zusammenarbeit, km = kilometer, O&M = operation and maintenance, OFID = OPEC Fund for International Development, OPEC = Organization of the Petroleum Exporting Countries, UGIIP = Urban Governance and Infrastructure Improvement (Sector) Project.

[a] Some *pourashavas* participated in more than one UGIIP.

Source: Authors.

Project Map

The Urban Governance and Infrastructure Improvement (Sector) Project has been implemented over three projects (2002–ongoing) in 96 *pourashavas*, about 30% of the country's 329 *pourashavas*. About 40% of the urban population resides in *pourashavas*. The central government ranks *pourashavas* by economic size: 193 are class A, 96 are class B, and 40 are class C.

Figure 3: Bangladesh—Urban Governance and Infrastructure Improvement Project *Pourashavas*

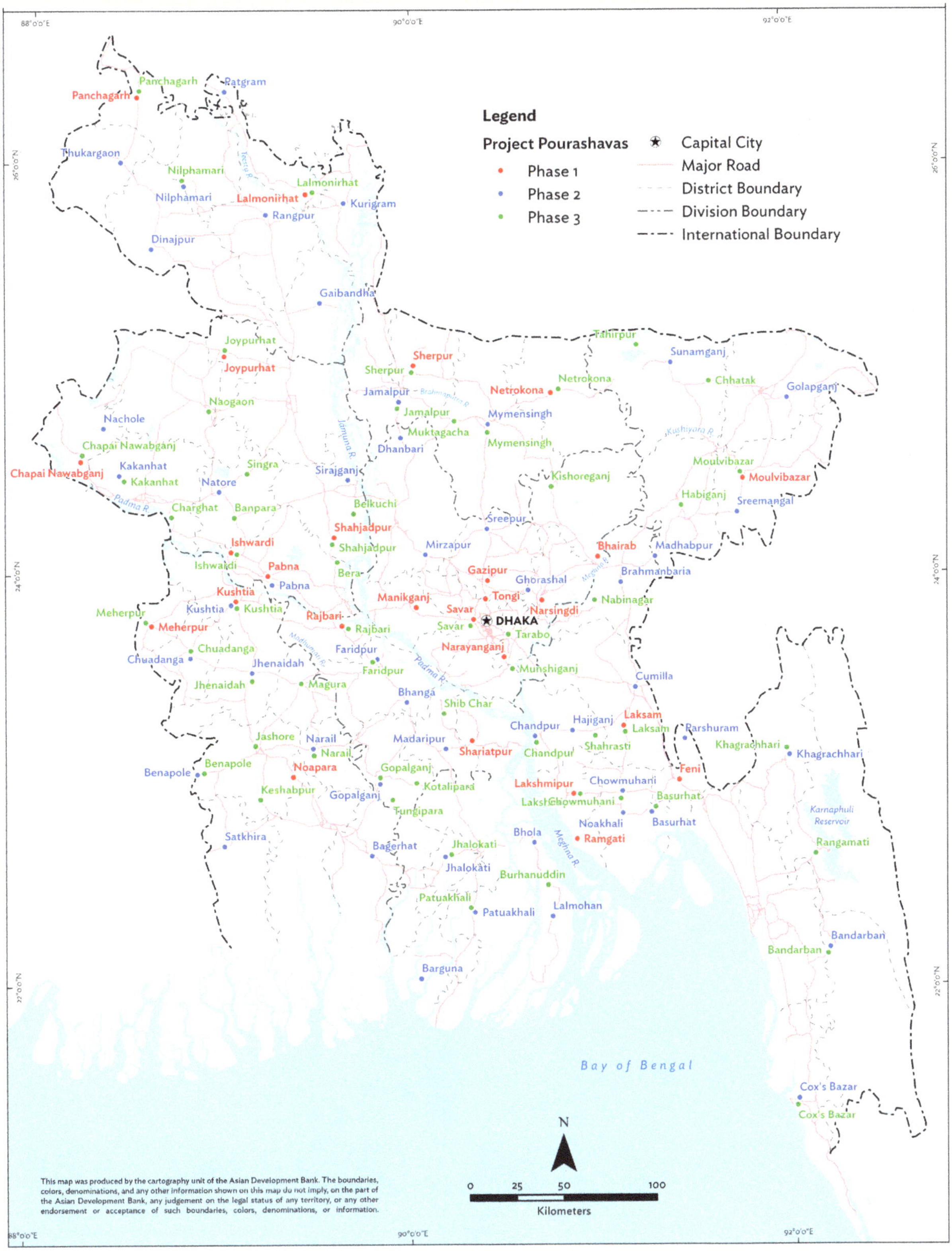

Source: Asian Development Bank.

Advocate for girls. Sharmin Sulatana, an elected councillor of Ghorashal *pourashava*, has used UGIIP to advocate for girls' education, and against child marriage.

INTRODUCTION

At the turn of the 21st century, nearly 30 years since Bangladesh's independence in 1971 and nearly as many years since the start of development assistance from the Asian Development Bank (ADB) in 1973, the country's urban areas seemed to be at a development impasse. The Government of Bangladesh and ADB were aspiring for more sustainable outcomes from urban projects, which up until that point had been traditional infrastructure-centric investments (assets) that cities were either unprepared to operate and maintain, or unable to leverage for new economic growth. Infrastructure alone was falling short of its potential in Bangladesh, yet infrastructure-driven development was the paradigm. Bangladesh's cities, however, needed a new approach, one that did not depend so singularly on infrastructure yet could still deliver things like water systems, roads, and sanitation—and, ultimately, economic momentum.

"It's true. There was no model for us. As an economist, the only tool I had was logic. I knew we needed results, not just assets," said Hun Kim, who conceptualized what became the 20-year-strong work of the Urban Governance and Infrastructure Improvement (Sector) Project (UGIIP). Kim is a retired director general of ADB's South Asia Department. "I was determined not to do the standard infrastructure packages. I thought to myself, I have only two choices: engage or not engage." The transaction costs for innovation can be high because of the time required to work out new, potentially better solutions.

Bangladesh's Local Government Engineering Department (LGED) and Kim's team worked from 1998 to 2000 on a consultative, deliberative process before official loan processing began. What they conceptualized was a new approach to urban development that has been quietly praised over the decades by government, mayors, and international development partners as a proven strategy for meaningful, transformative, prosperous urban development. "Urban development takes a long time. Don't take governance for granted," Kim said. "All urban development is governance work. Governments don't just want ADB for finance. They need us to build a support system, to build capacity."

> *"I knew we needed results not just assets... All urban development is governance work. Governments don't just want ADB for finance. They need us to build a support system, to build capacity."*
>
> **Hun Kim**
> UGIIP creator and former
> ADB director general of
> South Asia Department

A root cause of Bangladesh's urban development dilemma in the late 1990s was a lack of understanding the political economies of the country's *pourashavas*, the municipal governing bodies of the country's 329 municipalities of secondary cities and towns. The central government has maintained financial and administrative control over *pourashava* human and financial resources. The *pourashavas* themselves have not harnessed their inherent local potential (financially, politically, or socially) to fill the gap between the limited resources the central government has historically allocated to them and what they need to become more financially self-determined, self-sufficient, and prosperous. Residents, likewise, have seemed unaware of their *pourashavas'* latent potential to raise their own revenues to fund their own development, further compounding *pourashava* dysfunction with widespread lack of public expectation and participation. Poverty and economic growth rates in the 1990s suggest that the public was too preoccupied with daily survival to consider how civic participation might help improve their quality of life. But where there is public

Careful attention. Girls and women in Maulavibazar *Pourashava* receive sewing and specialized handicraft training as part of the ADB-supported UGIIP project to strengthen women's financial capacity.

expectation and participation, there can be accountability of leadership. This is what a diverse group of thinkers from government, civil society, and ADB came to believe was missing from the development equation in Bangladesh's *pourashavas*: governance.

In 2002, UGIIP was the first of three investment projects (UGIIP-1) to offer incentive-driven, performance-based funds that depended squarely on mayors leading a governance reform movement in their *pourashavas*. As participating *pourashavas* met best practice standards in good governance, they qualified for infrastructure loans, of which 30% had to be spent improving the living conditions of slum communities.[1] A *pourashava* would be disqualified from participating in the project if it failed to meet the designated criteria within the scheduled time frames. Being disqualified was a difficult lesson that only six *pourashavas* experienced, although they rejoined future UGIIP projects and succeeded.

With this publication, ADB and the government share their strategy of governance-driven urban development, with 20 years of experience in implementing UGIIP's evolving adaptations and results to improve its efficacy. Appendix 1 lists the key documents consulted for this report, and Appendix 2 lists the key people interviewed to supplement the documents and provide new insights. The report is divided into five chapters. Approached as a single narrative, this report begins by answering why UGIIP was needed then goes on to explain the concept of UGIIP, what UGIIP produced in terms of governance, capacity, and assets. The report takes stock of the many successes, challenges, and lessons before turning to what is needed next.

---

[1] Centre for Urban Studies, MEASURE Evaluation, and National Institute of Population Research and Training. 2006. *Slums of Urban Bangladesh: Mapping and Census, 2005*. Dhaka, Bangladesh, and Chapel Hill, North Carolina, United States, defines a slum as a neighborhood or residential area with a minimum of 10 households or a single unit with at least 25 members with one of the following conditions prevailing: predominately poor housing; very high population density and room crowding; very poor environmental services, particularly water and sanitation facilities; very low socioeconomic status for most residents; and lack of security of tenure.

Chapter 1, **"Context: Why Invest in *Pourashavas*,"** demonstrates the need for the governance strategy by examining the challenges *pourashavas* were experiencing at the onset of UGIIP-1. The economic hardship *pourashavas* went through and are still experiencing will likely be familiar to many mayors of secondary cities in Asia and the Pacific. UGIIP's governance-based development strategy should feel promising to these readers. **Chapter 2,** **"The UGIIP Strategy,"** focuses on the project-based solutions: how the UGIIP strategy has evolved to test new ideas and refine the approach. In the UGIIP governance reforms, mayors have the antidote to urban economic inertia, and that is better governance. Chapter 2 presents the eight governance criteria that *pourashavas* have had to progressively meet to continue participating in the project and accesses their project-allocated funds for infrastructure. Chapter 2 correlates the specific issues to each of the eight governance criteria to make their (i) basic infrastructure systems more sustainable, (ii) economies move, (iii) cities more livable, and (iv) communities more engaged in their own development. UGIIP-participating *pourashavas* typically far exceed the minimum mandatory achievements for each of the eight governance criteria. **Chapter 3,** **"The Latest Results: UGIIP-3,"** reports the general results in governance improvements and infrastructure improvements and illustrates the impact of those results—the participation, capacity development, and improvements in infrastructure and services—on people's lives as told through the "Impact Stories" that appear in this chapter (and also throughout the publication). In **Chapter 4,** **"Successes, Challenges, Lessons,"** the operational experience of ADB and the project staff, consultants, and contractors who implemented the projects provide what urban development professionals are looking for and what ADB seeks to share: the success factors, good practices, challenges, and lessons from the investment. The publication concludes with a brief **Chapter 5,** **"Unfinished Business,"** which shows that future iterations and adaptations of UGIIP can be made more effective by addressing some of the newer and more persistent challenges.

Development as a field of practice requires its practitioners, as well as the influencers and deciders, to examine enduring projects, such as UGIIP, for clues that unlock new ways of doing development. This knowledge product aims to both strengthen UGIIP as an emerging strategy as well as challenge other paradigms that are not delivering the impacts that cities in Asia and the Pacific need to address failing infrastructure, traction-less economies, and generational poverty. Past UGIIP project officers have stopped short of calling UGIIP a model. "I don't really believe in models," Kim said. "Every country is different. The capacity is different." And yet, UGIIP has influenced urban project designs in India, Sri Lanka, and Nepal, where variations of performance-based governance reforms have been integrated into largely infrastructure development projects. There are better ways of doing urban development, and UGIIP's governance-for-infrastructure strategy is one of them, bringing together the national and local body politic for economic and socially transformative development.

> *Pourashava*-in-Focus: **Naryanganj**

Impact Story 1: She Learned to Write Her Name with Project Group, Now She Does Its Banking

Rokeya Begum, 75, lives with her two sons and husband in a one-room tin house. At the doorstep, she has placed a cooking pot where she continuously makes cakes to sell in the market. But the cakes are only a sweet side business.

"I can earn Tk1,000–Tk1,500 daily (about $12–$17) selling cakes. Business was good in the beginning, so I took another loan and bought a van rickshaw," Begum said. "Now I have eight van rickshaws. My income in 2022 has increased 10 times."

Begum started her enterprise in 2010 from a *pourashava* microfinance program started as part of the ADB-supported UGIIP. "In 2010, I could afford to eat once in 3 days. I could not buy new clothes for many years. We had no means to survive," Begum said. "Then UGIIP implemented its project in this extremely poor place of ours."

Begum's community, Dhokhin Rally Bagan, is a low-income neighborhood of Narayangonj City Corporation, which participated in UGIIP when it was still classified as a *pourashava*. "We lived in dirt and garbage, then the *pourashava* constructed a footpath and drainage line in 2011 after UGIIP started."

Under its third term elected mayor, Selina Hayat Ivy, the city corporation has maintained a microfinance fund since 2011 for residents with limited financial or social capital for escaping poverty. The decade-long fund began as a UGIIP initiative, but has been sustained by municipal funds. "This is, I believe, one of our biggest achievements," the mayor said.

The women of Dhokhin Rally Bagan, about 255 in total, have organized themselves into cooperatives of about 15 women each that participate in training and manage microcredit funds, all begun under UGIIP, but sustained by the city.

Roma Datta, a community worker from Narayanganj City Corporation, has worked for the project since its earliest interventions. "There are 17 groups, and each group has 15 women members. All of them received loan facilities, and they are also handing their collective money through government banks," Datta said. "They are financially literate and doing business from their savings accounts."

Begum smiles at how much she has learned from participating in her local women's cooperative. "I learned how to write my name from the group session. Now I am the leader for the bank transactions," she said. "This made me so proud."

Women advocates. Rokeya Begum, left, an entrepreneurial woman from the low-income community of Dhokhin Rally Bagan, speaks with Roma Datta, right, a community worker from Narayanganj City Corporation, who has worked for the project since its earliest interventions.

Source: Authors.

INTRODUCTION

Champion of female financial literacy.
Since the earliest days of the project, Roma Datta, a community worker from Narayanganj City Corporation, has helped organize and train women to make business sense out of their livelihood skills. Read more on page 5.

CHAPTER 1:
WHY INVEST IN POURASHAVAS

Bangladesh is a rapidly urbanizing country, but different types of cities in the country are urbanizing at different rates. In addition, a few metropolitan areas are urbanizing at an extremely fast pace compared to the rest. These two trends cast an urgent spotlight on the country's secondary towns—*pourashavas*—where a considerable proportion of the urban population is settled, for now, offsetting further population pressures on the megacities.

Working against the prospects of *pourashavas* has been unpredictable fiscal decentralization, resulting in a lack of any legal mechanisms to determine how fiscal resources are distributed to local governments. Policy makers in Bangladesh understand that achieving upper middle-income country status by 2030 is not achievable without effective, equitable fiscal decentralization. For that to happen, municipalities need good urban governance, characterized by competitive performance standards and public accountability.

Growth rates. For decades, the growth rates of urban populations have been twice the national rates.[2] Since 1991, the urban population growth rate hovered around 3.5%[3] before slowing to 1.3% in 2019 (footnote 2). The total urban population in 2020 was estimated at 62.8 million, or 38.17% of the total population (footnote 2). The government expects the country's urban population to reach 74 million, or 41.6% of the total population, by 2025.[4]

Urban centers. There are 506 urban centers in the country (Table 1). According to the government's 2020 comparative analysis of urbanization in Bangladesh and other Asian countries, Bangladesh showed the highest rate of urban primacy at 32% (the urban population share of primate cities, such as Dhaka and Chattogram[5]), with Viet Nam coming in a distant second at 23.2% and Pakistan at 22.5% (footnote 2). Bangladesh also has the fewest cities with more than 1 million people—just 3 compared to 6 in Viet Nam and 10 in Pakistan. This makes for an urban population dispersed over many smaller cities.

Table 1: Distribution of 506 Urban Centers in Bangladesh

Type of Urban Center	Population Range by Urban Center Type	Number of Urban Centers by Type
Rural service centers	<25,000	240
Subregional centers	25000–99,999	220
Regional centers	100,000–499,999	39
Metropolitan Areas (Chattogram, Khulna, Rajshahi)	500,000–4.9 million	3
Dhaka Metropolitan Area	>5 million	1

Source: Government of Bangladesh, Planning Commission. 2020. *Eighth Five Year Plan, July 2020–June 2025*. Dhaka. pp. 595–596.

---

[2] Bangladesh Bureau of Statistics Population Census, 2011 (Analytical Report, 2015) as cited in Government of Bangladesh, Planning Commission. 2020. *Eighth Five Year Plan, July 2020–June 2025*. Dhaka. p. 594.
[3] World Bank. Data by Country. http://data.worldbank.org/indicator/SP.POP.TOTL (accessed 5 January 2022).
[4] United Nations. 2015. *World Urbanization Prospects: The 2014 Revision*. New York.
[5] Formerly Chittagong; its name was officially changed in 2018 to better reflect its Bengali pronunciation. This publication uses Chattogram.

The individual population size of the smaller cities should not mask their importance. Collectively, these 500-plus secondary cities and small towns represent more than 53% of the country's total urban population. Most local governing bodies (329) are *pourashavas*.[6] The hierarchy of urban local government is explained in Table 2. Although most of this growth has been in the country's primate cities, the population growth of *pourashavas* has also consistently exceeded the national average. About 40% of the total urban population in Bangladesh reside in *pourashavas*. These *pourashavas* act as a buffer—an alternative destination—between the rural areas and the already overwhelmed larger cities.[7] Urban land area is also relatively limited to just 11,125 square kilometers (km^2), just 7% of the country's land mass, creating high urban density. In 2011, the average density of urban areas was 4,028 people per km^2 compared to 790 people in rural areas.

Table 2: Local Government by Type

Type of Urban Local Government	Type of Urban Area
City Corporation (12)	Largest cities (Dhaka North, Dhaka South, Chattogram, Khulna, Rajshahi, Sylhet, Barisal, Rangpur, Mymensingh, Naryanganj, Gazipur, and Cumilla)
Pourashava (329)	Annual revenue of Tk2–Tk8 million
Class A	Annual income of at least Tk10 million and at least 75% holding tax collection rate for the last 3 years (without government transfer or grant)
Class B	Annual income of at least Tk6 million and at least 75% holding tax collection rate for the last 3 years (without government transfer or grant)
Class C	Annual income of at least Tk2 million for the last 3 years (without governments transfer or grant)

Tk = taka.
Source: Government of Bangladesh, Local Government Division. 2011. *Circular of 31 May 2011*. Dhaka.

Pourashava Challenges

Urban growth has been more of a challenge than an opportunity for *pourashavas*. Investing in them, however, can act as a jetty against the tide of rural migration to the country's megacities, offering some relief to their already burdened infrastructure. However, *pourashavas* have been places of unfulfilled potential—a liminal place between countryside and big city, where infrastructure has aged and lagged spatial growth.

Pourashavas have been overwhelmed by several issues common to unruly urbanization, stagnant development, and the particular political histories of government administration in the country.

---

[6] *Pourashavas* are local government institutions incorporated under the Local Government (*Pourashavas*) Act 2009. There are four conditions to be met to be a *pourashava*: (i) three-fourths of the people are involved in a non-agricultural profession, (ii) 33% of the area must be non-agricultural land, (iii) it must have an average density of no fewer than 1,500 people per km^2, and (iv) its population is not less than 50,000.

[7] Asian Development Bank (ADB). 2012. *The Urban Governance and Infrastructure Improvement Project in Bangladesh: Sharing Knowledge on Community-Driven Development*. Manila. p. 5.

Urban challenges have been even more acute in *pourashavas* because of severe fiscal weakness, capacity constraints, and a governance status quo that is disconnected from the public over municipal business, which is critical to keeping officials accountable. The most tangible evidence of these challenges has been chronic infrastructure deficiencies and poor public services in *pourashavas*.

Urban Infrastructure and Services

Rapid urbanization has created a growing demand for urban infrastructure and services, and poor urban governance has created a gap between the inability to supply and the pent-up demand for infrastructure and services. Even where urban infrastructure is improving, it is still not keeping pace with the population growth rate or the cost for operation and maintenance (O&M) of infrastructure systems. *Pourashavas* are struggling to provide basic services such as drinking water, sanitation, roads, drainage, solid waste management, and other urban amenities that matter to everyday quality of life, such as kitchen markets, streetlights, parks and open spaces, and bus terminals.

Through a series of three projects of the Urban Governance and Infrastructure Improvement (Sector) Project (UGIIP), the Asian Development Bank (ADB) has supported improvements, first in governance and subsequently in *pourashava* infrastructure.[8] *Pourashavas* still need significant investment to enhance municipal infrastructure and service delivery, to strengthen urban climate change resilience and reduce regional disparities across the country. The following municipal services have yet to experience transformational levels of investing, though 96 of the country's 329 *pourashavas* are in a substantially stronger governance position to responsibly access infrastructure financing and project implementation through their participation in UGIIP. All dates, unless otherwise stated in footnotes, are based on the Government of Bangladesh's Eighth Five Year Plan, 2020–2025.

- **Water supply coverage.**[9] Since 2000, water supply coverage in the *pourashavas* has improved in real numbers, though the percentage of access to piped water supply has remained steady at about one-third of households because of the population growth rates in *pourashavas*. Piped water systems are only available in one-third of *pourashavas*, typically only for a limited time, about 2–4 hours per day. The water quality is typically poor, largely with high iron content, high salinity in coastal areas, and sometimes arsenic contamination.
- **Sewerage coverage.** Except for a limited sewerage system in Dhaka, no other urban center had any form of sewerage system in 2000, when UGIIP-1 was being conceptualized. In 2015, 58% of the urban population had access to improved sanitation facilities.[10] Fecal sludge management remains a major challenge, especially in urban slums.

---

[8] UGIIP-1 (2002–2010) provided a loan of $65 million to support 27 *pourashavas*. UGIIP-2 (2008–2016) provided a loan of $87 million and cofinancing of $40.8 million equivalent to support 51 *pourashavas*. UGIIP-3 (2017–ongoing) provided a loan of $100 million to support 35 *pourashavas*. ADB also provided additional financing of $200 million to expand UGIIP-3 to up to 20 additional *pourashavas*. Major infrastructure achievements under the three projects include rehabilitation or new construction of 2,265 km of urban roads, 724 km of drains, 224 km of piped water supply and 60,000 individual meters, 27 solid waste disposal sites, and 82 municipal facilities.

[9] The National Policy for Safe Water Supply and Sanitation (1998) assigns the *pourashavas* the responsibility of urban water supply (except in the three hill districts and the divisional towns).

[10] World Health Organization. 2015. *Progress on Drinking Water and Sanitation: 2015 Update*. New York.

- **Solid waste.** Solid waste collection levels have stagnated at 20% without any systematic service. Households have few options than to dump their solid waste directly into streets, public spaces, and drains, which may protect their households from immediate health and hygiene issues, but can cause broader public health risks and environmental problems.
- **Housing.** Most urban residents live in precarious shelters. Only 25.73% of urban houses are made of permanent materials (bricks or cement) (footnote 2). Unaffordable land prices, insecure tenure, and a lack of housing finance have stunted the formal housing market.
- **Drainage.** Drainage is often underdeveloped and poorly maintained. During monsoon rains, roads and pathways often flood, causing severe traffic congestion and risks to public health.
- **Roads and bridges.** Many urban roads and bridges are dilapidated because of high traffic volumes, street congestion, and lack of maintenance. Frequent traffic jams in city centers limit economic productivity and access to social services and everyday destinations, such as schools, hospitals, banks, and markets.
- **Environmental quality.** In most urban centers, ambient air and water quality is extremely poor. The major sources of air pollution are traffic, brickyards, and industry. The major sources of water pollution are domestic and industrial wastewater discharges. Inadequate drainage increases the exposure of the urban population to the effects of polluted water.
- **Climate vulnerabilities.** Major climate risks for *pourashavas* include rising temperatures, higher intensity and frequency of rainfall and wind loads, storm surges, and riverine flooding. Rising sea levels may affect *pourashavas* in coastal zones and intrude inland, increasing the salinization of both groundwater and surface water. *Pourashavas* need enhanced institutional abilities to sustain O&M of infrastructure and services, and to work with poor and vulnerable communities to reduce their exposure to risks and increase their resilience to the effects of gradual climate change as well as calamities.

Poverty and Inequality

Bangladesh has experienced a steady decline in national, urban, and rural poverty rates since 2000, though the coronavirus disease (COVID-19) pandemic has reversed some of those gains and has temporarily threatened progress toward national goals. When UGIIP-1 was being conceptualized in 2000, the national poverty rate was 49% (comprising 62.8 million people). By 2016, this had been cut in half to 24% (comprising 39.1 million people) (footnote 2).

During that same period, rural poverty rates fell faster than urban poverty rates, while extreme rural poverty rate fell at twice the pace as extreme urban poverty. The principal reason behind this phenomenon is large-scale migration from rural to urban areas. The pace of poverty reduction slowed from 2010 to 2020 because of slower growth in the job market, which had fueled the gross domestic product per capita gains of 2000–2010.

Inequality among households is acute in urban areas, with the rich living in central business districts and the poor in urban slums. In its Eighth Five Year Plan, the Government of Bangladesh acknowledged that the rapid economic growth of the country since 1990 has not been inclusive. The degree of inequality as measured by the Gini coefficient increased from an average of 0.38 in the 1980s to 0.46 from 2000–2009, with inequality highest (and worsening) among urban households (0.47) than rural households (0.43).

Financial power in organizing. The women of Dhokhin Rally Bagan, about 255 in total, have organized themselves into cooperatives of about 15 women each for livelihood trainings and management of their own microcredit accounts, all begun under UGIIP, but sustained by the city.

Despite significant progress in poverty reduction, a large portion of the urban population still lives below the poverty line, and the economic growth is bypassing them. In 2016, the monthly urban household income was about 70% higher than rural household income levels (footnote 2). Economic opportunities in urban areas will continue to attract large numbers of migrants from rural areas. Without decent housing or basic services, the next waves of migrants will join multiple generations of migrants that still make their homes in slum areas. These trends point to the continued need to address urban poverty, especially in the wake of the pandemic's impact on employment income and the exposure of cities to inadequate infrastructure to protect public health.

Community Participation

Many *pourashavas* have not had the institutional arrangements (e.g., committees, forums) in place for communities to collectively express their needs, preferences, and ideas to elected representatives and local officials. Municipal leadership is also working from an inherited history that did not involve the public in discussions about policies under consideration, budget development, and service delivery. As a result, top-down decisions have not benefited poor communities as much as needed. This gap between public service and public needs has meant that elected officials are not held accountable for their decisions and results, or lack of results, during their tenure. A lack of transparency has become the standard condition of government operations and what the public has come to expect. The latent governance of *pourashavas* is most evident in the lack of key urban infrastructure and services: water supply, sanitation, drainage, and solid waste management. The Local Government (*Pourashava*) Act of 2009 has initiated reforms and generated improvements, primarily through town-level coordinating committees (TLCCs) and ward-level coordinating committee (WLCCs), which built on the successful experience of TLCCs and WLCCs under UGIIP-1. Women and the poor have been gaining visibility and a voice.

Pourashava-in-Focus: **Benapole**

Impact Story 2: Child Marriages Decline with Project-Inspired Community Advocacy

In 2011, Khaleda Khatun noticed positive change in Namazgram, a slum community in Benapole *Pourashava*. She began to see it in the environment, and then the voices of women—one of those being hers. "Women had no idea how they could contribute to their communities," Khatun said. "Child marriage, domestic violence, and abuse were part of our daily life before we started with UGIIP."

The first change that came for Namazgram was a newly constructed road as part of UGIIP. Before the new road, residents had to walk barefoot through mud to get anywhere. There were also no toilets or tube wells in the community. Without any streetlights, people could not be outside safely at night.

The new road established a higher baseline for mobility and cleanliness in the community. The road literally paved the way for what came next: a tube well, streetlights, trash bins, and hygienic toilets. Without the road, these other improvements were not possible or practical. More than 900 residents of Namazgram have benefited from these improvements. Female residents formed small groups to take care of their areas.

Khatun represents the Namazgram community on the TLCC, and organizes meetings to plan more improvements—and not just for infrastructure. The domestic lives and safety of girls and women are priorities for Khatun and others. "In our monthly meetings, we discuss women's empowerment. Many women from the slum participate in training programs organized by the *pourashava* under the GAP," Khatun said.

Khatun is working to prevent child marriages and establish a school in the community. She was married at a very young age, and though her husband supported her in completing her schooling, Khatun said her fortunate situation is not what often happens with most young brides.

"Many lives of women and children have been shattered because people are not aware of the consequences of child marriage and dowry," Khatun said. The occasional practice of committing one's child to marriage and offering financial incentives in matchmaking is a complicated and often secretive practice tied to economic security, especially among low-income and vulnerable households. Other UGIIP-participating *pourashavas* have taken on this social taboo as part of their community development.

"In this area, since UGIIP started," Khatun said, "not a single girl has gotten married before she turned 18. This is a huge achievement of our community."

Woman and child advocate. The governance reformers have given women such as Khaleda Khatun opportunities to emerge as local leaders. Khatun raises awareness about the harmfulness of child marriage and dowry practices in her low-income community of Namazgram in Benapole *Pourashava*.

Source: Authors.

Capacity Gaps

Urban governance has lacked resident participation, accountability, and sound fiscal management. *Pourashavas* are governed by an elected mayor and councilors. Mayors have traditionally dominated decision-making. Until UGIIP, residents had few opportunities to participate in the decisions that would or should affect them. Md. Shafiqul Islam Akand, former UGIIP project director (2011–2015) and retired additional chief engineer for the Local Government Engineering Department (LGED), points to capacity constraints as an underlying reason for the disappointing outcomes (mostly the sustainability) of urban infrastructure projects in the 1990s. "There was very low capacity among people who worked in municipal government and a lack of ownership of infrastructure and resources for O&M," Shafiqul said. "Infrastructure was built by the central government, therefore, it belonged to them and was their responsibility whether it functioned or not."

Technical capacity. *Pourashavas* in general are managed by many untrained staff, and approved local posts are often left vacant, pending appointments from the central level because *pourashavas* lack the legal authority to hire locally.

Legal capacity. The governance of *pourashavas* has been constrained by inadequate legal authority. Since 1977, *pourashavas* had been governed by a municipal ordinance. Over time, the ordinance proved inadequate for the autonomy that *pourashavas* needed to decide and make manifest their unique destinies provided by geography, history, people, and natural resources. *Pourashavas* needed what other countries and states have in official urban policy or in a lawfully binding act. Later, the Local Government (*Pourashava*) Act of 2009 was enacted to clarify the legal basis of *pourashavas* and their functions.

Planning capacity. Unplanned and rapid urbanization in *pourashavas* has created large unmet demand for urban infrastructure and services. Most *pourashavas* do not have a land-use or infrastructure development plan. Only class A *pourashavas* have an urban planner in the approved organizational chart, but in the past, the position has often remained vacant for a long time because *pourashavas* have a difficult time attracting and retaining qualified candidates, and they compete with larger markets where their services are in high demand. The country also does not have a long history of educational training in the field of urban planning to have a large pool of professional urban planners. Limited human resources are linked to poor planning and the precarious financial situation of *pourashavas*. Unplanned development creates an even greater strain on inefficient allocations of what limited funds are available for specific purposes. Under this prevailing weak governance structure, *pourashavas* hobble along without the incentives, knowledge, or capacity of elected leaders and *pourashava* officials to generate own financial resources for investment, staffing, and meeting public needs.

Financial capacity. *Pourashavas* have not had the capacity to generate sufficient revenues and so depend on budgetary transfers from the central Government of Bangladesh, which account for more than half of the total revenues for most *pourashavas*. The *pourashavas* have not had the systems in place for effective tax collection that would have helped generate revenues for essential infrastructure and services. From 2000 to 2002, when UGIIP-1 was being conceptualized and designed, only 29% of *pourashavas* collected more than half of their estimated holding tax, their most important revenue source.[11]

Fiscal Decentralization

For *pourashava* development to have a significant chance of reform, effective fiscal decentralization is required as a prerequisite. The low level of *pourashavas'* contributions to both revenues and expenditures indicates the effectiveness of the country's fiscal decentralization. The most recent data on local government finances is from fiscal year 2013, but as the Eighth Five Year Plan states, trends from fiscal year (FY) 1974 to FY2013 are indicative enough because "nothing has changed in terms of fiscal decentralization."

The structure of *pourashavas* revenue sources in 2013 was 43% from central government transfers, 43% from local charges and fees, and 14% from local property taxes. Still, local government revenues continue to account for less than 4% of total government revenues. On the expense side, local government spending in 2013 accounted for only 8.2% of total government spending, representing only 1.1% of gross domestic product. From 1974 to 2013, local government spending increased only 7%.

Government transfers. The share of government transfers had reached a high of 62% before *pourashavas* registered a 122% increase in own-source (local) revenues during 2011–2013. A peculiar challenge in Bangladesh is the lack of a legal mechanism for the central government transfers that establishes a formula for allocations or the timing. "The resource transfers are determined centrally based on political considerations and competing national priorities," states the Eighth Five Year Plan (footnote 2, p. 512). Some progress has been made for the lowest category of local governments, whose government transfers are now based on a formula that includes population and performance.

Taxes. The trend in taxation and cost recovery (tariffs) from public services has been negative, from 5.1% of total revenues in 1974 to 2.54% in 2011 before increasing to 3.82% in 2013. Holding (property) taxes are the single most important local revenue source that local governments have any control over. Although the central government determines tax rates and property valuation, local governments are responsible for their tax collection rates. "Yet, yields are insignificant," the Eighth Five Year Plan states (footnote 2, p. 504), "mostly owing to the lack of political will to design and enforce a well-defined holding tax. There are problems with land records, property valuation (mostly done on historical prices and costs), tax assessments, very low tax rates, and very weak tax administration."

Tariffs. Improving the tax collection system of *pourashavas* is a logical place to start with reforms and capacity building, but strengthening cost recovery has proven to be more reliable. Taxes as a share of local government revenues decreased from a high of 75% in FY1974 to 43% in FY2013, while the falling share of government transfers has incentivized local governments to correct its fees and charges for public services.

---

[11] World Bank. 2007. Bangladesh: Strategy for Sustained Growth. *Bangladesh Development Series.* No. 18. Washington, DC.

An international comparative study of centralization for the Eighth Five Year Plan showed that Bangladesh is among the most centralized countries in the world in terms of both expenditures and taxes.[12] "Good city governance and greater fiscal autonomy tend to be positively correlated and together they enable better services," the report stated (footnote 2, p. 516).

A Chance for Change

For all the challenges that *pourashavas* embody, ADB and the government believed that *pourashavas* should be valued for the opportunities they represent—opportunities to reduce poverty, improve quality of urban living, and support environmentally smart economic growth. But something needed to change in how ADB and the government was going about *pourashava* development. The government had long recognized the power of well-planned urbanization, and that good governance is a critical factor of successful urbanization. Effective poverty reduction is closely linked to effective urbanization strategies, such as public administration reforms, anti-corruption measures, decentralization, and the strengthening of accountability and participation. Furthermore, good governance is recognized as a critical factor encouraging economic growth and poverty reduction (Box 1).

Box 1: Defining Governance

When the Urban Governance and Infrastructure Improvement (Sector) Project (UGIIP) was first presented to the Asian Development Bank (ADB) Board of Directors for approval in 2002, the project's designers were proposing a new approach to infrastructure development. The project designers proposed that project towns would have to earn their access to infrastructure financing by improving their governance practices. To ensure a broad understanding of governance, the project team took precious space in the bank's highly regulated template to define and explain "governance."

The project proposal to the Board of Director's borrowed UN-Habitat's definition of good urban governance: "Urban governance is the sum of the many ways individuals and institutions, public and private, plan and manage the common affairs of the city."[a]

The proposal went on to explain two aspects of the broad definition. "First, it indicates that urban governance is not just government. Governance as a concept recognizes that decision-making power to improve the affairs of the city exists inside and outside the formal authority and institutions of government. In other words, governance includes the government, private sector, and civil society. The second aspect emphasizes that urban governance is a 'process.' It is a participatory process through which diverse interests of the city may be accommodated and cooperative action can be taken."[b]

[a] UN-Habitat. Global Campaign on Urban Governance: Principles.
[b] ADB. 2002. *Report and Recommendation of the President to the Board of Directors on a Proposed Loan and Technical Assistance Grant to the People's Republic of Bangladesh for the Urban Governance and Infrastructure Improvement (Sector) Project*. Manila.

Source: Authors.

---

[12] Bhal, Linn, and Wetzel. 2013. *Financing Metropolitan Governments in Developing Countries*. Lincoln Institute of Land Policy. Cambridge, Massachusetts.

In 2001, ADB realized that it needed a new approach to urban development in Bangladesh. Prior to the approval of UGIIP-1, ADB had invested in five other projects for water supply and urban development, starting in 1982. Until that point, ADB had rated its implementation of physical works under past and ongoing projects *satisfactory*, but the institutional reforms that are integral to the sustainability of those physical works were not gaining traction. ADB diagnosed the slow institutional reform as a symptom of the poor governance and lack of accountability afflicting the project-participating municipalities. Cities cannot function well without good governance and administrative management.

"A popular modality at the time was to set up a fund and take applications from cities," explained Hun Kim, the first mission leader who conceptualized UGIIP and a former director general of ADB's South Asia Department. "But I thought, no way will it work. And with a fund, you are just a banker. In Bangladesh, you can't be a banker. You have to be a development worker." Kim believes local knowledge is where ADB is strong, and something new could be figured out. "All the work comes from resident mission staff. They are the logic behind a lot of projects," he said. One asset ADB had going to for it, Kim said, was the highly capable LGED—they know what will work and what will not.

There was another yellow flashing light cautioning ADB from moving forward with business as usual in Bangladesh's cities. ADB had provided technical assistance to the government to prepare the Third Urban Development Project for Secondary Towns.[13] The project was shaping up to be more of the same, until ADB was met with significant delays in processing the proposed loan project. Nearly 3 years had passed without any progress. For this reason, ADB decided to reformulate the project and learned that governance should be the primary focus.[14] The people of *pourashavas*—the mayor, the council, the residents—needed a reorientation to all aspects of governing—planning, power sharing, revenue generation, investing in people, and community-centered development. The forthcoming project would introduce a new performance-based approach. Participating municipalities would be made accountable for their performance in implementing action-based programs for better urban governance and "rewarded" with access to finance for urban infrastructure.

---

[13] ADB. 1997. *Technical Assistance to Bangladesh for the Third Urban Development Project*. Manila.
[14] ADB. 2001. Proposed Loan and Technical Assistance Grant to the People's Republic of Bangladesh for the Urban Governance and Infrastructure Improvement (Sector) Project. Manila.

Pourashava-in-Focus: **Khagrachari**

Impact Story 3: Housing Brings "Different Kind of Strength" to Person with Disabilities, Widows, Veterans

For the first time in her 20-year-old life, Sazi Akter Sazu feels surrounded and supported by people who can relate to just how difficult life has been for her. As a child, Sazu contracted typhoid and lost her ability to walk and finish school. "All my life, I went through humiliation as a person with disabilities," Sazu said.

Her prospects began to change when Khagrachari *Pourashava* allocated special infrastructure improvement funds to building homes for families living under extreme duress—those whose members include people who are differently abled, widowed, landless, or elderly veterans. "It did not take a lot of time for me to feel at home," Sazu said. "Every person I know in this *abashon* (community) has a story. Their life is like mine, so I do not feel left out. We united and found our peace in solidarity. It would never have happened if we had not gotten the chance to live together."

Khagrachari *Pourashava* successfully implemented its reforms and allocated its allotted infrastructure funds to building the Bangabandhu Pouro Abashon Prokolpo (Bangabandhu Municipal Housing Project). The community is made up of 33 one-storey homes with two bedrooms, a living room, kitchen, and toilet. Each home has a water connection and uses cylinder gas for cooking. Residents use collective funds for paying their utility bills. A community garden and play area are popular gathering places.

"This housing from the municipality and UGIIP has given me a new identity. I consider myself a human being now," Sazu said. "Many disabled people live in houses near me, and together, we are working hard to change our fate. This house is like a dream. If I did not have this home, my daughter and I would end up on the road. Now I can look forward to doing something with our lives, and this is a different kind of strength."

Sazu has a 6-month-old daughter, and is studying from home to finish her secondary education. "Now I dream about my daughter. I want her to get the best education," Sazu said. "This house has become a reason for me to live and dream for a better tomorrow."

Finding community. Khagrachari *Pourashava* allocated some of its UGIIP infrastructure improvement funds to building homes for families living under extreme duress—those whose members include people who are differently abled, widowed, landless, or elderly veterans.

continued on next page

Impact Story 3 *continued*

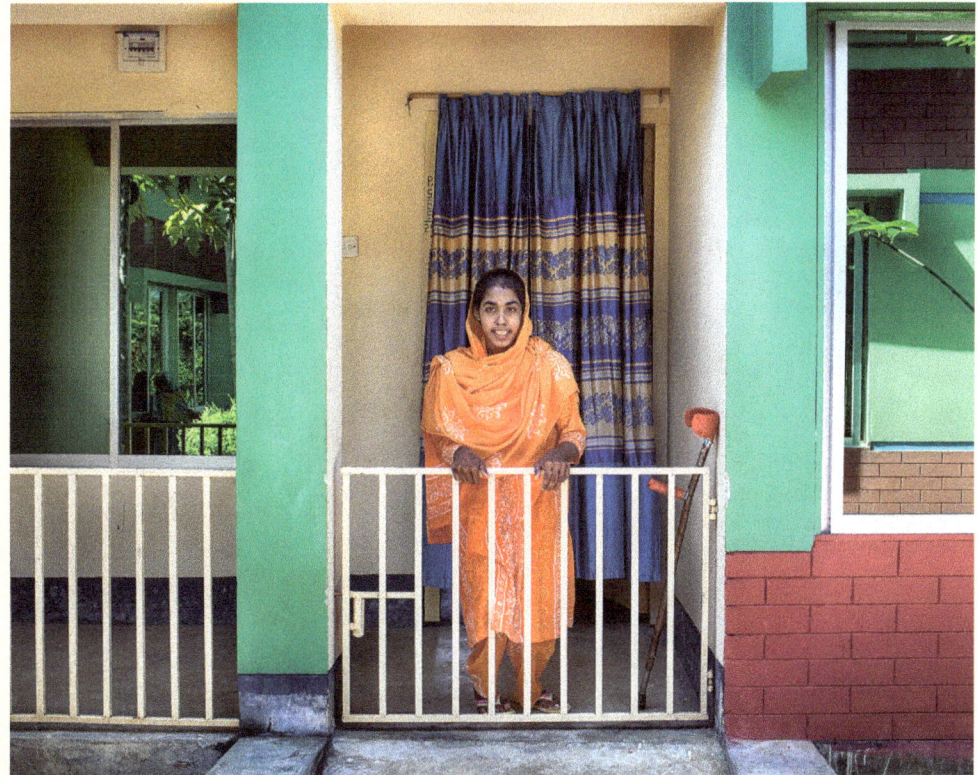

Solidarity among neighbors. Sazi Akter Sazu, 20, has difficulty walking because of a childhood case of typhoid. She found community and security as a resident of the UGIIP-supported housing project for residents who are differently abled, widowed, landless, or elderly veterans.

Investing in the youngest residents. With housing secure, families of the Bangabandhu Municipal Housing Project can shift their aspirations to their children's education, which is supported by a new community school and outdoor study space for children.

Source: Authors.

Not the same place anymore. Jhorna Begum and 60 other families are the residents of Durga Narayan Pur Dhuli Bari, a low-income community in Sherpur *Pourashava* that was once characterized by slum conditions, but is now making gains with project investments in water, sanitation, paved roads, street lighting, and more.

CHAPTER 2:
THE UGIIP STRATEGY

The Urban Governance and Infrastructure Improvement (Sector) Project (UGIIP) challenges the status quo approach to urban infrastructure improvement projects. The experience of the Asian Development Bank (ADB) has shown that infrastructure cannot be sustained unless the underlying governance problems are addressed first. How to address the governance problems, and which governance problems should be addressed, comprised the development challenge for Bangladesh.

The opportunity to redesign the Third Urban Development Project for Secondary Towns helped ADB and the government, in close consultation with *pourashavas*, identify the common governance problems of *pourashavas* and an agreeable—though innovative—way of addressing them and the infrastructure needs. The preparation of UGIIP-1 followed a consultation process that spanned nearly 2 years and included academicians, journalists, and nongovernment organizations.

This chapter looks at how the UGIIP strategy was conceptualized, beginning with the first project. The chapter also presents the strategy's centerpiece, around which governance requirements and infrastructure revolve. The chapter also explains how *pourashavas* were prioritized and selected for each of the three UGIIPs out of the 329 *pourashavas* nationwide. ADB and the Local Government Engineering Department (LGED) tweaked the design of UGIIP-2 and UGIIP-3 based on their experience implementing UGIIP-1.

Project Preparation

Two general and significant concerns emerged across *pourashavas*. First was the concentration of authority and powers in the *pourashava* mayor.[15] This singular hegemonic role faced little public accountability, which partly explains *pourashavas'* poor financial and service performance and inefficiency: no one was holding anyone responsible. The second concern was the weak capacity of elected commissioners, local public officials, and community groups. Leaders at all levels need the skills for taking on new roles that are necessary for decentralized responsibilities and greater local autonomy. As roles and responsibilities increase for municipal councils, ward councils, and various resident committees, their capacity must also increase. From these two general findings, the design team behind UGIIP-1 knew the project needed to accomplish three things: the project needed to increase (i) delegation, (ii) participation, and (iii) accountability within *pourashavas*. Good local governance is not possible without all three of these interdependent forces working together.

Decentralization and Delegation

Decentralization of authority and powers from the central government to *pourashavas* is futile if the process stops at the mayor's office. Mayors must also delegate powers to the elected commissioners and to resident groups. Without delegation, the mayor's office becomes a bottleneck—too narrow of a channel to move progress and opportunities through—and the intentions of decentralization, i.e., more effective government, is not as possible. Delegation rarely happens naturally or voluntarily, though, as it is not a historical political tradition in many cities in the developing world. Delegation needs incentives.

---

[15] The Local Government (*Pourashava*) Act of 2009 changed the title of the elected head of *pourashavas* to "mayor" from "chairperson".

Participation and Accountability

By delegating responsibilities to the elected commissioners and resident groups, their participation develops and creates an ecosystem of accountability (   ). Once the public is participating more in the decision-making processes of their *pourashava* and community, they have an interest—a more personal stake—in whether the government's efforts are succeeding or not. The public's engagement and awareness of what the municipal government should be doing is how elected officials feel accountable for their performance and initiate changes when the public is not satisfied. Accountability becomes a natural function of government when there is meaningful participation of residents in *pourashava* affairs.

These key factors for successful decentralization and local governance—delegation, accountability, and participation—have been lacking in *pourashavas*. The project design team (comprised of consultants, the LGED, and ADB, in close consultation with *pourashavas*) understood that it needed the project to, first, incentivize mayors to delegate. And what does the public want most? Basic infrastructure. What ADB and the LGED had learned from past project experience, though, is that the *pourashavas* need more than infrastructure. *Pourashavas* need what basic infrastructure needs: good governance.

"If we talk about 'urban development,' infrastructure is just one part of the story while good governance is the other essential part of running a city," said Alexandra Vogl, the mission leader for UGIIP-3's additional financing and current principal planning and policy specialist in the Strategy, Policy, and Business Process Division of ADB's Strategy, Policy, and Partnerships Department. "For me as an urban planner, UGIIP has all of the essential components of a comprehensive urban development project."

> ### Box 2: The Role of Central Government in Local Government Affairs
>
> While urban local bodies are ultimately responsible for urban governance, the central government has important influence and control over various aspects that determine local autonomy. Fiscally, the central government finances its own municipal projects and has statutory control over certain aspects of municipal revenues and expenditures. The central-level Ministry of Local Government, Rural Development and Co-operatives is responsible for supervision of municipal affairs, the scope of which includes demanding records, inspecting activities, and revoking illegal orders and decisions of municipal governments.
>
> Given the large number of the central government agencies involved in planning, building, and managing urban infrastructure, the first Asian Development Bank-financed Urban Governance and Infrastructure Improvement (Sector) Project (2002–2010) reviewed the functions and structure of the central government ministries and agencies in urban development. The focus of the review was based on the principle that the role of central government agencies should be more of facilitator or supporter than of doer or implementer for the sake of long-term benefits of the local governments.
>
> Source: Authors.

From their consultations, the UGIIP-1 designers identified and rallied consensus around five basic reforms: (i) resident awareness and participation, (ii) women's participation, (iii) integration of the urban poor, (iv) financial accountability and sustainability, and (v) administrative transparency. It was expected that these reforms would exponentially improve any *pourashava's* daily operations, the functionality of its infrastructure and assets, and its sustainability. The designers made these reforms conditions to accessing financing for basic infrastructure improvements.

Money for infrastructure doesn't come easily or often for *pourashavas*. They rely on central government budget allocations (transfers), government loans, and property taxes. Transfers are not enough to cover their basic operating expenses, which puts *pourashavas* in debt (many in arrears on loans) for capital projects.[16] *Pourashava* property tax collections tend to be based on old assessments, and collection levels are not what they should be. *Pourashavas* in general were also not innovative or were too politically timid to take advantage of other potential sources of local revenues—non-holding taxes, fees, rental charges, and leases on public property and equipment.

Without dedicated leadership to these reforms, *pourashavas* would not be able to achieve them. The UGIIP-1 designers required ambitious and willing mayors. Without these qualities, they risked failure and being disqualified from the project and any chance at infrastructure finance. No reform, no infrastructure. The UGIIP-1 designers were clear on this strict stipulation. "It's not efficient to suspend or allow municipalities to drop out," said Masayuki Tachiiri, the mission leader for UGIIP-2 and current director of the Strategy, Policy, and Business Process Division of ADB's Strategy, Policy, and Partnerships Department. "That's why we spent so much money on capacity building and support for municipalities to meet the targets."

For a more in-depth read of one *pourashava* mayor's experience implementing UGIIP, see Appendix 6.

---

[16] According to the Eighth Five Year Plan, local governments are prohibited from market borrowing. The central government is the source of all grants (mostly for *pourashavas*) and loans (mostly for larger city corporations) in the form of transfers. The central government does not expect local governments to have the capacity to repay loans, thus, the loans stay on the books as outstanding dues.

Pourashava-in-Focus: **Benapole**

Impact Story 4: Abuse Survivor Turns Award-Winning Entrepreneur

The training center in Benapole *Pourashava* accommodates more than 30 girls and women who are learning handicrafts and sewing. The trainees visibly admire their trainer, Shuli Rani Dey, 38, a role model for many women who have experienced domestic violence and other abuse.

"When my abusive husband fled town 8 years ago and left the burden of his loan on my shoulder, I was devastated," Dey said. She and her 7-year-old son survived by working in people's houses. "We had not one kilogram of rice at home. I washed people's cloth for Tk10 (about $0.13)," Dey said.

During her hardship in 2014, Dey heard about the training program offered by UGIIP-3.

Dey enrolled in the training program, and after 3 months took out a small loan to purchase a sewing machine and began taking orders from her neighbors. "That was my turning point, and I quickly started earning money. From Tk500 monthly income as a housemaid (about $5), I started earning Tk3,000–Tk5,000 a month (about $29 to $48)," Dey said. "UGIIP-3 project saved my child's life and our future."

She dreams of having a shop once she completes the reconstruction of her decade-old house. "I recently renovated my house and purchased some gold jewelry for myself as an investment. But my main asset is my son and his education," Dey said. Her son is now attending college.

She emphasized how the UGIIP-3 training program helped her provide for her son and his education. "It's unfortunate, but most young people here get involved in the workforce by leaving education. I am fortunate that I can continue my son's education," Dey said.

Dey began training at the *pourashava* training center because she wanted to give back to her community, help other women facing hardship (which she can relate to), and guide young women so they can earn by themselves. On International Women's Day in 2019, the Department of Urban Development awarded Dey the "Shafal Nari Award, 2019."

Since 2013 the UGIIP-project has supported the training of more than 800 women and girls. "Women and girls have advanced their careers and started earning independently," Dey said. "There is a big market for local handicrafts, and we are trying to emerge in the workforce by learning and investing."

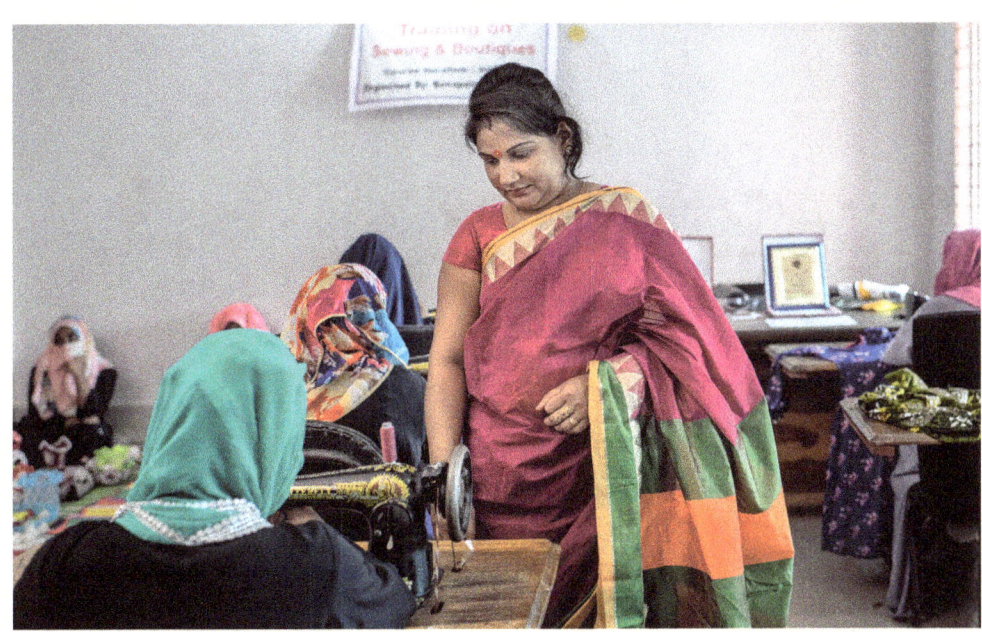

Trainee turned trainer. Shuli Rani Dey, 38, a role model for many women who have experienced domestic violence and other abuse, has trained more than 800 women in her *pourashava* on how to sew and run their own small business.

Source: Authors.

Project Design

UGIIP may be best understood as offering *pourashavas* three general opportunities: (i) financial incentives to achieve meaningful reforms; (ii) technical assistance and capacity building to implement the reforms; and (iii) the earned, performance-based financing for infrastructure that communities value and will also invest in. The chance to earn infrastructure funds was the incentive mechanism for reforms and a strategy for shared growth. "We thought people would not accept this, but, in the end, people understood the concept as something for their betterment," said Md. Shafiqul Islam Akand, former project director (2011–2015) and former additional chief engineer for the LGED. "Three things converged under UGIIP: gender development, participatory government, and the self-reliance of municipalities."

Reforms: The Centerpiece of Transformational Change

To develop *pourashavas'* capacity for delegation, participation, and accountability, the original project design team developed the Urban Governance Improvement Action Program (UGIAP).

The UGIAP is the centerpiece of the UGIIP strategy and has been refined by each of the three UGIIPs. UGIIP-1 conducted consultations for 2 years to develop the first UGIAP, which identified five governance areas as the starting point to developing good governance for the majority of *pourashavas*. UGIIP-2 added an additional indicator focused on better physical planning to control development, and UGIIP-3 expanded the reform program with two more indicators focused on raising local revenues and keeping essential *pourashava* services functional. The UGIAP now covers seven key areas of good governance with as many as 28 sub-activities (Figure 4). *Pourashavas* who have participated in more than one UGIIP (some have participated in all three) are evaluated against more stringent indicators.

> *"Three things converged under UGIIP: gender development, participatory government, and the self-reliance of municipalities."*
>
> **Md. Shafiqul Islam Akand**
> UGIIP project director, 2011–2015, and former additional chief engineer for the LGED

"Municipalities entering UGIIP were sick bodies—not functioning at all, but they got healthy and self-reliant by going through the UGIIP reforms," Akand said.

Figure 4: Seven Areas of the Governance Improvement Action Plan under the Third Urban Governance and Infrastructure Improvement (Sector) Project

Source: Urban Governance and Infrastructure Improvement (Sector) Project Management Office.

Reform Area 1: Residents' Awareness and Participation in Municipal Decision-Making

The UGIIP strategy is based on the logic that before *pourashava* officials can be expected to be responsive and accountable, they must first have a clear understanding of what residents want and residents must understand the roles, responsibilities, and capacity of their municipality. This area of reform organizes residents into officially recognized committees and introduces decentralized urban planning, which is further elaborated on in Reform Area 5. It implements systems and mechanisms for communicating with residents (e.g., public boards in each ward, published notices, semiannual leaflets, grievance-redressal cells in government) and provides communities ways to voice their needs and opinions of government priorities, decisions, and overall performance.

To improve residents' participation in official municipal decision-making, all participating UGIIP *pourashavas* have been required to establish town-level coordinating committees (TLCCs) with 33% female members and ward-level coordinating committees (WLCCs) with 40% female members. Before UGIIP-1, only 5 of the 27 project towns had any comparable kind of civilian committee.

To enhance public awareness and accountability, a strategy for greater transparency has required *pourashavas* to publicly post the previous year's budget outcomes that compare between public budget proposals, final budgets, and actual budget outlays.

By UGIIP-3, this reform area had become focused on establishing the committees and using the established participatory budget processes and communication channels to develop resident charters and report cards on municipal performance—the original intent of this reform area.

Reform Area 2: Equity and Inclusiveness of Women and the Poor

Every project *pourashava* has had to establish two standing committees: one for women and children's affairs and a second for poverty reduction and slum improvement. These two committees must each have 40% female representation. The municipality, TLCCs, WLCCs, and other standing committees work together to develop, approve, and implement three distinct social development plans: a gender action plan (GAP), a PRAP, and a slum improvement plan. According ADB's Hun Kim, UGIIP-1 was the first ADB project to implement a GAP, which had been newly operationalized at ADB.

Before UGIIP-1, only 5 of the 36 project towns had developed any kind of comparable plan. These plans form parts of the *pourashava* development plan (PDP). Since UGIIP-2, the project *pourashavas* have been required to self-fund these plans. At least 1% of the municipal revenue had to be earmarked for implementing the *pourashava*-specific GAP, of which 50% had to be spent by the end of the project.

"UGIIP required the preparation of a GAP and that municipalities start to spend some of their own revenues on implementing this plan," said Saito, UGIIP-3 mission leader and current director of ADB's South Asia Urban and Water Division. "Before this project, the mayors and the engineers in the *pourashavas* had limited exposure to how a municipality could support gender development, such as through women's employment and skills improvement. But by having the GAP as a requirement, they saw real changes. They saw the benefits of supporting gender equality and poverty reduction and the GAPs. Seeing this is how the process can be sustained."

UGIIP-1 also included the establishment of a microcredit fund with a utilization plan and livelihood training for poor women. By UGIIP-2, the municipalities had taken over the management of the fund and activities. Based on the impact stories throughout this publication, microcredit programs are still operational and have been replicated by women at the community level.

> *"Before this project, the mayors and the engineers in the pourashavas had limited exposure to how a municipality could support gender development..."*
>
> **Norio Saito**
> UGIIP-3 mission leader and current director of ADB's South Asia Urban and Water Division

Reform Area 3: Financial Accountability and Sustainability

Two committees established by UGIIP, and now legally mandated through the Local Government (*Pourashava*) Act of 2009, are the Standing Committee on Establishment and Finance and the Standing Committee on Accounts and Audit. The mandatory involvement of these two financial committees would bring greater transparency and participation into the preparation of *pourashava* budgets and the mandatory account audits.

Computerization also makes accountability, transparency, and sustainability more possible. Each project town has been required to computerize its tax records and accounting records for computerized billing, collections, and reporting. Computerization, coupled with current property tax appraisals and regular interim tax assessments and reassessments, was meant to support the requirement for *pourashavas* to increase their collection of holding taxes. Towns were expected to achieve 85% holding tax collection efficiency. At the beginning of UGIIP-1, only 5 of the 27 project towns had such efficiency levels. Towns were also required to increase revenues from local nontax sources, which often needed to be identified. Where nontax revenues were being collected, *pourashavas* were required to increase them at least in accordance with inflation.

To remain eligible for the project, *pourashavas* had to remain current on all electricity and telephone bills and remain on schedule with their debt repayment to the central government and any other lender; many of these accounts were overdue and in deep arrears. This meant *pourashavas* had to generate revenues to pay these bills and remain qualified to stay in the project, which included, for instance, the ability to access their allotment of infrastructure spending.

"We knew that payment of bills, including those in arrears, was a big issue in other projects," said Masayuki Tachiiri, the mission leader for UGIIP-2 and current director of the Strategy, Policy, and Business Process Division of ADB's Strategy, Policy, and Partnerships Department. "It is a conscious effort of ADB as a whole to look after the opportunities and impacts to other utilities in our projects."

Reform Area 4: Administrative Transparency

Good urban governance depends on efficiency, and for that, *pourashavas* were required to map and match their responsibilities and needs with the appropriate staff, which under UGIIP, would receive the necessary training and support of subcommittees. *Pourashavas* have been required to develop a clear staff structure with detailed job descriptions. Over time, UGIIP required municipal governments to also introduce e-governance and information technology into their processes. Training programs have been extended from only elected officials to all staff and residents concerned with *pourashava* affairs.

Reform Area 5: Urban Planning

UGIIP-2 added to the UGIAP framework by creating a new set of reform indicators on physical planning with the ultimate expectation that *pourashavas* would be able to control development. Class A *pourashavas* have been required to recruit a full-time urban planner, though this has proven to be difficult. Physical planning begins with a verified base map and updated land use plan, which *pourashavas* have had to prepare as inputs to the required PDP, which like all plans UGIIP required, had to use gender-inclusive processes. The PDP must include an annual O&M plan of municipal assets with budget requirements. The O&M budget is approved as part of the approval for the PDP. (As noted below in Reform Area 7, in UGIIP-3, O&M was also elevated because of its importance to sustainability and good governance, and given to its own reform area.)

Improved planning has also required *pourashavas* to complete an inventory of public assets and produce an endorsed municipal infrastructure development plan. These assessments are fundamental to improving several municipal functions, such as financial planning for O&M, rehabilitation, and expansion of infrastructure.

Reform Area 6: Enhancement of Local Resource Mobilization

UGIIP-3 culled from Reform Area 3 those indicators related to revenue generation and elevated them to their own reform area. This change in the UGIAP brought a more singular focus to each finance-related reform area. Reform Area 3 under UGIIP-3 was refocused on the preparation of an annual *pourashava* budget; the auditing of accounts; computerizing of the accounting system, ensuring payment of all bills and loan schedules; and a new indicator related to fixed assets. *Pourashavas* under UGIIP-3 have been required to complete an inventory of all fixed assets, open a fixed asset register, design a fixed asset database, and create a fixed asset depreciation fund account. In addition to computerizing the tax and accounting systems (including billing and collection) to increase tax and nontax revenues, under Reform Area 6, *pourashavas* were required to establish and collect a water tariff. This involved *pourashavas* investing (where applicable) in water meters to bill and collect via computerized and bank payments volume-based tariffs with an 80% collection efficiency.

Reform Area 7: Keeping Essential *Pourashava* Services Functional

UGIIP-3 added Reform Area 7, which is about operating, maintaining, and sustaining public services and amenities, such as keeping streetlamps lit and public transportation terminals and lavatories clean and in good working condition. To ensure the sustainability of *pourashava* assets and services, each UGIIP-3 town has had to prepare and approve an annual O&M plan with resources allocated from its own budget. The funding for the O&M plans is made possible by the successful implementation of the reforms, such as increased tax collection and other new revenue streams. By adding this seventh reform area, the UGIIP-3 designers also hoped to send a signal to non-project *pourashavas* of the importance of keeping local assets and services in good working condition.

Infrastructure: Performance-Based Financing

Pourashavas that meet the governance requirements receive their allotted allocations for infrastructure improvements. Municipalities that did not meet the requirements within the given time frame are disqualified from participating further in the project. In the history of UGIIP, only four *pourashavas* have been dropped from the project. By linking reforms to financial disbursements, the project motivates *pourashavas* to improve governance while also providing them with tangible development impacts from the infrastructure investments.

Pourashavas could use UGIIP infrastructure financing on the following six types of infrastructure projects: (i) slum improvement, (ii) urban roads and drains, (iii) water supply, (iv) sanitation, (v) solid waste management, and (vi) municipal facilities. At least 5% of infrastructure funds had to be spent on slum improvements.

Funding arrangements. UGIIP-3 disbursements to *pourashavas* for infrastructure subprojects are organized into three phases of governance reforms: (i) 20%–25% of the disbursements in the initial phase of building awareness and participation, (ii) 60% in the intermediate phase focused on improving urban planning, and (iii) 15%–20% in the advanced phase concentrated on sustainability. The UGIAP indicators were grouped according to the phase they most relate to, e.g., the phase 1 indicators included establishing the TLCCs, WLCCs, and standing committees. These tended to be the most achievable governance indicators and ones that established the accountability and participation aspects of the project.

Each phase was linked to a specific set of criteria for each reform area (Appendix 4 and Appendix 5). For each phase of the project, *pourashavas* were required to achieve the minimum performance ratings within the allotted time frame. Otherwise, they became ineligible to move forward to the next phase(s)

Pourashava-in-Focus: **Benapole**

Impact Story 5: From Stay-at-Home Mom to Salon Owner

Sanjida Sreeti Ranu's personal life and fortunes have intersected with an ADB-supported project in more than one way. In January 2020, Ranu, 32—along with 50 other women—took a 3-month training course in salon services ("beautification"), offered by UGIIP-3. A year later, she was the owner of a salon in the Noor Shopping Complex. "If I had to pay to take this course, I would never have been able to join," Ranu said.

Ranu's husband is employed but unable to earn enough, while she is also responsible for taking care of their two young sons and her widowed mother most of the time, when she could be earning money outside the home. "I had no idea how to start a business," Ranu said. "Once I had my training, I gained the confidence to take a small loan and start my business. Within a year, I was earning more than my husband."

Her business also benefited from UGIIP-3 infrastructure improvements to the main road in front of the Noor Shopping Complex. Where there were once open drains, footpaths and closed drainage have been built. Businesses are thriving, including Ranu's Sanjida Beauty Parlor. "Businesses are doing well because people are no longer hesitant to come to this road," Ranu said. "I pay Tk4,000 (about $47) rent for this space. But every month my income is Tk20,000 (about $232) and sometimes more. Because of the good road condition and footpath, things have changed. I cannot imagine what would have happened to my business if the road had not been built."

The Noor Shopping Complex hosts 50 shops that stand to gain from the more accessible road conditions. "People were not very welcoming in the beginning. But once they saw how I worked hard and thrived with my parlor, people started appreciating it," Ranu said. "I am not only contributing to my family, but I have also become a part of this business community now. This gives me hope to do more."

Business sense. Sanjida Sreeti Ranu opened a salon after completing a livelihood and business training program.

Good for business. A UGIIP-supported road rehabilitation in front of Sanjida Sreeti Ranu brought her more customers.

Source: Authors.

and receive additional investment allocation. Achieving a fully satisfactory performance in phase 2 entitled a *pourashava* to the balance of its funding allocation and made it eligible for additional funds for infrastructure investments from the unused allocations of *pourashavas* that did not achieve the minimum or achieved only the minimum performance in their phase 2 milestones.

Selection of Participating *Pourashavas*

UGIIP has been or is being implemented in 96 of the country's 329 *pourashavas*. Each of the UGIIPs increased the number of *pourashavas* covered. Though a recurring recommendation from each project has been to reduce the number of *pourashavas* per project to increase the impact on the poor and improve project manageability, demand has been too high for ADB project designers and loan processing officers to reduce the number of participating *pourashavas*. Nevertheless, the investment level has increased significantly per project.

UGIIP-1. The first UGIIP selected 27 *pourashavas* considered by the central government to be high-priority investments. They had also received earlier support through the ADB-financed Secondary Towns Infrastructure Development Project I and the World Bank-financed Municipal Services Project.

UGIIP-2. UGIIP-2 was implemented in 35 *pourashavas* (29 class A and class B towns, and 6 class C towns). UGIIP-1 towns were not eligible for UGIIP-2. The six class C towns represented one from each of the six administrative divisions trying to achieve some regional balance among the 29 class A and B towns; the project assessed and ranked them according to objective indicators based on their weighted scores. These indicators included demographic features, economic potential, infrastructure deficiency, investments received in recent years, incidence of poverty, management and administrative capacity of *pourashavas*, and creditworthiness. For each indicator, sub-indicators were developed, and appropriate weightage was given to measure the *pourashava*'s score against these sub-parameters.

Currency fluctuations in the taka created significant savings that enabled the expansion of UGIIP-2 to include 16 more *pourashavas*. Although UGIIP-2 was unable to offer the same funding amounts to these additional *pourashavas*, their inclusion in the project was still a boon for them.

> *"By the time UGIIP-2 came, mayors were already aware of UGIIP and had the incentive to join and do some governance work even before getting any money."*
>
> **Masayuki Tachiiri**
> UGIIP-2 mission leader and current director of ADB's Strategy, Policy, and Business Process Division

UGIIP-3. To develop "model towns" for other *pourashavas* to aspire to, UGIIP-3 supports *pourashavas* through two schemes. The first scheme ("window A") provides infrastructure funds and capacity development support for 36 high-growth potential *pourashavas* with a total population of 2.2 million. The second scheme ("window B") allocates up to $2 million in infrastructure improvement funding to each of the top 20 *pourashavas* in a governance assessment. Window B is based on well-defined performance criteria. The implied competitive nature of the assessment made it possible to influence a larger number of *pourashavas* in the area of governance improvement using the UGIIP principles. The 36 window A *pourashavas* were selected based on population size, density, and growth; own source revenues; level of past investments; status as a district headquarters; the 20 window B *pourashavas* underwent a compliance assessment with intermediate UGIAP criteria (see Appendix 5 for the intermediate-phase governance indicators).

Evolution of Design

Over the course of 20 years of implementing the three UGIIPs, each project improved upon the previous approach and tried new ways to accommodate the demand for more *pourashava* participation. In UGIIP-1, governance was the key problem. UGIIP-2 further addressed this challenge by expanding the UGIAP with indicators related to urban planning and staffing. In UGIIP-3, urban climate resilience has emerged as a new focus area.

Changes from UGIIP-1 to UGIIP-2. A major contribution of UGIIP-2 was the reorientation of the development process to begin at the ward level with "visioning exercises" to inform higher local government levels of the wards' development aspirations and priorities—where they believe their taxes and the city's revenues should be spent. The PDP developed from the consolidation of this visionary thinking from wards.

UGIIP-2 adopted a different disbursement approach, scheduling the first payment to become eligible only after each *pourashava* had established its standing committees and finalized its *pourashava* development plan. "By the time UGIIP-2 came, mayors were already aware of UGIIP and had the incentive to join and do some governance work even before getting any money," said Masayuki Tachiiri, UGIIP-2 mission leader and the current director of the Strategy, Policy, and Business Process Division of ADB's Strategy, Policy, and Partnerships Department. "They knew they would get their allocations if they did well, but that wasn't the case in UGIIP-1. Mayors were naturally skeptical then."

By delaying initial disbursement and offering only limited funds initially, Tachiiri said the capacity and seriousness of *pourashavas* were tested. One mayor recalled, "For the first 2 years, we did not see any UGIIP funding, and that delay created a negative motivation that we had to work against." Mayors had an additional incentive with UGIIP-2: larger funding packages.

The disbursement is worth a *pourashava's* patience and diligence, since for many of the participating *pourashavas*, UGIIP is the largest opportunity for infrastructure funding that they have ever had access to.

Changes with UGIIP-3. To motivate *pourashavas*, UGIIP-3 has used a strategy of building momentum by introducing entry criteria that have enabled *pourashavas* to become eligible for 15%–20% of infrastructure funding, and using easier governance criteria for the first 2 years.

With high demand to participate in UGIIP, the third project adopted a new structure to allow more *pourashavas* with various capacity levels and experience with reforms to participate. UGIIP-3 introduced two windows that a *pourashava* might gain entry to the project:

(i) **Window A,** the primary scheme, provides infrastructure funds and capacity development support for governance improvement to 36 preselected *pourashavas* with a total population of 2.2 million.
(ii) **Window B,** the secondary scheme, allocates funds according to well-defined performance criteria on a competitive basis to all remaining *pourashavas* in the country. The top 20 qualifying *pourashavas* are allocated up to $2 million in financing for infrastructure improvement. Window B *pourashavas* are measured against a separate set of UGIAP criteria that uses objective and measurable governance indicators.

The UGIIP-3 design introduced higher infrastructure funding levels for window A *pourashavas* and added the O&M reform indicator. UGIIP-1 and UGIIP-2 offered a maximum of $1 million to $2 million per *pourashava*. "Although improvements were happening," said Saito, of the UGIIP-3 team, "they were still piecemeal. A road is improved here, drainage is improved there, a slum is improved, but they were not being improved in an integrated, holistic way. So, we wanted to increase the amount of funding."

Pourashava funding maximums went from $1 million to $10 million (or more, in some cases), depending on the *pourashava's* size and capacity. Even at this level, infrastructure needs and finance demands are still not met. "The improvements need to be done gradually and in tandem with the capacity enhancement at the *pourashava* level," said Norio Saito, UGIIP-3 mission leader and ADB director of ADB's South Asia Urban and Water Division. "If our support to each *pourashava* is too high—beyond the capacity of the *pourashava*—in the end, they can't manage the assets. So there is a tradeoff."

Appendix 3 shows the changes to the UGIAP over the three projects, expanding from five reform indicators for UGIIP-1 to eight indicators for UGIIP-3.

Supporting Policy and Inspiring New Policy

At the international level, UGIIPs were designed to support the country's progress toward Sustainable Development Goal 11: to make cities and human settlements inclusive, safe, resilient, and sustainable. As in past UGIIPs, UGIIP-3 was designed to support the government's Seventh Five Year Plan, 2016–2020, which identified urban development as a national priority, specifically, the reduction of urban poverty and the improvement of living conditions through better governance and services. UGIIP was also designed to be fully consistent with ADB's Urban Operational Plan, addressing inclusiveness, environmental sustainability, and competitiveness; and the Bangladesh country partnership strategy (2016–2020), which prioritizes inclusive and environmentally sustainable growth, governance improvement, and climate- and disaster-resilient infrastructure and services.[17] The significant improvement in urban development and management inspired several national institutional and sector reforms, including the Local Government (*Pourashava*) Act of 2009 and the draft national urban policy. The next chapters look at the results and experiences of implementing UGIIP as designed.

---

[17] ADB. 2012. Urban Operational Plan 2012–2020. Manila. Available at https://www.adb.org/sites/default/files/institutional-document/33812/files/urban-operational-plan-2012-2020.pdf.

Pourashava-in-Focus: **Bhairab**

Impact Story 6: Mother of Five Builds Successful Shoe Business with $60 Loan

Chandila Tila, a minority community in Bhairab *Pourashava* in central Bangladesh, is known for its communal livelihood: making or repairing shoes and leather crafts, among other low-wage work. Homes often double as workshops, and both men and women of the household are engaged in the craft.

Kanchi Rani Das, 35, is a mother of five children, living in Chandila Tila, where her husband's family has lived for two generations before him. About 1,500 people live in Chandila Tila. "Before UGIIP-2 started in our area, we were living in filth… packed with people and livestock," Das said. "We had no footpath or drainage facility."

First came a new road and drainage near Chandila Tila, then sewing and other handicraft training for low-income women. "In one group, 15 to 20 people learned different handicrafts," Das said. "Besides that, I received a Tk5,000 loan (about $60) to continue a shoemaking business with my husband, who is a cobbler." They pooled their money with others to rent a shop and collectively repair and sell new shoes. Now, she earns Tk15,000 monthly (about $175).

Das is also a leader of a "loan community"—where women contribute to a savings account, which lends to women to expand their businesses. "From UGIIP, we have learned how women can also have savings," Das said. "There are women in my community whose husbands left them or died, and now they are supporting their children alone."

Having never attended school, Das taught herself how to read and write, and now, two of her children go to a school supported by Bhairab *Pourashava* revenues generated from the UGIIP reforms. "We had almost no educational facilities in the past. Now, our children go to school for free," Das said. She dreams of sending her children to university.
Emotional at recalling how much her life has changed for the better, Das wiped her face with her blue cotton sari, and she showed off a one-bedroom house filled with homey decorations and new furniture. "I bought the majority of my stuff with my savings," she said. "I could never imagine this before I started my work."

The cobbler's wife. Kanchi Rani Das, 35, took her husband's cobbler business and increased its profits beyond her wildest expectations after completing a project-sponsored livelihood training program on business management.

Source: Authors.

Sweet dreams. Rakeya Begumis says she has significantly increased her daily earnings by selling homemade cakes, using entrepreneurial skills that she and 255 women learned as part of a women's group organized in 2006 by the UGIIP-1 project, with ADB support.

CHAPTER 3:
THE LATEST RESULTS

The results of the first Urban Governance and Infrastructure Improvement (Sector) Project (UGIIP) and UGIIP-2 have been reported extensively in various publications over the years, but with UGIIP-3 nearing completion, the time was ripe to inventory the results and impacts of the third project in this series. At the time of this publication, UGIIP-3 had conducted four annual benefit monitoring and evaluation (BME) surveys to determine the extent the project was achieving the intended benefits. The BME surveys cover progress and benefits of the new various standing committees, new infrastructure construction, and improvements in *pourashava* services.[18]

This chapter summarizes the results of UGIIP-3 as it nears completion. The impacts of these results are illustrated in corresponding case studies that anecdotally tell how individuals have benefited from UGIIP-initiated activities. The presentation of results is organized according to the two main investment areas of UGIIP: (i) improvements in *pourashava* governance and (ii) improvements in *pourashava* infrastructure and services (Table 3).

Aside from addressing immediate needs of individual *pourashavas*, UGIIP-3 contributes to the national and global efforts to meet the Sustainable Development Goals. Table 4 identifies the variety of Sustainable Development Goals addressed in the design and implementation of UGIIP-3.

Table 3: Project Investment Areas

Governance	Infrastructure
Resident awareness and participation	Roads and culverts
Urban planning	Drains
Equity and inclusiveness of urban poor	Solid waste management
Enhancement of local resource mobilization	Water supply
Financial management, accountability, and sustainability	Basic services to the urban poor
Administrative transparency	Municipal facilities
Keeping essential *pourashava* services functional	Operations and maintenance

Source: Authors.

---

[18] Government of Bangladesh, Local Government Engineering Department (LGED); and Government of Bangladesh, Department of Public Health Engineering. 2022. *Our Talks: Success Stories of Self-Reliant Women in UGIIP-III Under LGED*. Dhaka. p. 4.

Table 4: Project Contributions to the Sustainable Development Goals

Sustainable Development Goal Area	Contribution by UGIIP-3 Investment Area
No poverty	Basic services to the urban poor
Zero hunger	Basic services to the urban poor
Good health and well-being	Drains, solid waste management, water supply, sanitation, basic services to the urban poor
Quality education	Municipal facilities
Gender equality	Roads, municipal facilities
Clean water and sanitation	Water supply, sanitation, basic services to the urban poor
Decent work and economic growth	Basic services to the urban poor
Industry, innovation, and infrastructure	Drains
Reduce inequality	Roads, basic services to the urban poor, municipal facilities
Sustainable cities and communities	Drains, solid waste management, water supply, sanitation, municipal facilities
Climate action	Roads, drains, water supply, sanitation, municipal facilities

SDG = Sustainable Development Goal, UGIIP = Urban Governance and Infrastructure Improvement (Sector) Project.
Source: Asian Development Bank.

Cleaner cities. Magura *pourashava* developed a fecal sludge treatment plant (FSTP) with a capacity of 5 m^3 per day, using funds it earned from governance reforms through UGIIP-3, supported by ADB.

Pourashava-in-Focus: **Chapainawabganj**

Impact Story 7: Water Connection Ends Rushing, Queuing, Joint Pains

Nur Jahan Begum spent most of her life worrying about securing basic needs, such as water supply. After more than 70 years, Begum received her first household water connection. Her days of queuing twice a day at the community tube well 2 miles away from home are over.

"I never had enough time to do things. I was always rushing to queue with my water pot," she said. "When I got a water connection for the first time, it was something very special to celebrate. Getting water any time of the day that you need is a different kind of privilege. I think I am very lucky." Begum said she suffered from joint pain, but now feels relief from not having to queue for water anymore.

In Notun Para, 58 households received individual water connections, transforming the area. Residents say it is cleaner, and people can maintain better hygiene. Women in the area have also participated in the project-supported livelihood training. "Before we got this connection, we were collecting water from tube wells. During summer, we survived simply with little water," Begum said. "Now we get water inside our kitchen. This was a long-awaited dream for the community."

She and her three sons live in a three-room ancestral house. Two of her sons are married, and Begum said she never wanted an impoverished life for her daughters-in-law and grandchildren. "Our children missed their play times to collect wood or collect water," Begum said. "Now I feel very happy that my grandchildren will not miss class or leave school for collecting water for the family."

Waiting for water. Nur Jahan Begum says her life changed when her ancestral home received its first household water connection—which delivers a clean, metered, and affordable supply on-demand to her home.

Source: Authors.

Governance Improvements

UGIIP departs from conventional urban development projects, as it prioritizes improvements in urban governance and overall capacity raising ahead of infrastructure development. The major benefits expected from the reform actions were improved revenue collection, resulting in improved local services and local development. To build the capacity of project *pourashavas*, UGIIP-3 supported trainings on key local development topics and supported the shift to modern computerized accounting, billing, tax assessment, and communication (similar to UGIIP-1 and UGIIP-2). An estimated 37,781 people from 36 *pourashavas* participated in training. Participants included mayors, councilors, representatives of low-income communities, women, civil society members, and members of the *pourashava* staff and standing committees. This section looks more closely at improvements in *pourashavas'* financial capacity, service capacity, and residents' awareness of and participation in standing committees.

Financial Capacity

UGIIP has incentivized *pourashavas* to improve their inefficiency rates on property tax collection, other taxes, and fees (such as water tariffs), and to discover other local revenue resources, making it possible for the *pourashavas* to pay their bills, repay loans, and reduce their dependence on central government budget allocations (transfers).

"Municipalities did not know how to explore their own potential for resources. As a result of participating in UGIIP-3, most *pourashavas* have increased their revenues by at least double," said Md. Shafiqul Islam Akand, former project director (2011–2015) and former additional chief engineer for the LGED. UGIIP-3 *pourashavas* have often exceeded the 85% target increase in holding tax collection efficiency and 40% increase in non-holding tax collection, which are revenues from service fees they had not been collecting. Since UGIIP-3 began, all project *pourashavas* have nearly doubled their demand for holding and non-holding taxes, from $6 million–$11 million before the project to $13 million–$22 million. Collection efficiency before UGIIP-3 ranged from 30%–40%. Now, more than 80% of all holding taxes are successfully collected and almost 100% of non-holding taxes are successfully collected.

The results of building *pourashava* capacity for revenue collection have been impressive since UGIIP-1. Tachiiri recalled from his time preparing UGIIP-2, "I was very happy when I saw the *pourashavas'* performance with revenue collection. They collected revenues in good ways, and some collected more than what they got from UGIIP. So, they started using their money for better services and that made mayors even more popular. There is huge potential for revenue generation in the *pourashavas*."

The new revenues enabled *pourashavas* to achieve the mandatory UGIIP reforms of paying overdue bills, remaining current on all electricity and telephone bills, and remaining on schedule with debt servicing; many of these accounts had been overdue and in deep arrears. Table 5 summarizes the significant improvements in *pourashavas'* capacity to repay their bills and remain current.

Table 5: Payment Capacity of UGIIP-3 *Pourashavas*

Accounts Payable	Payment Efficiency Rate Before UGIIP-3	Current Payment Efficiency Rate
Electricity bills	Below 40%	More than 90%
Telephone bills	Below 80%	100%
Government loans	Below 50%	More than 80%

UGIIP = Urban Governance and Infrastructure Improvement (Sector) Project.
Source: UGIIP Project Management Office. 2020. *UGIIP-3 Annual Report 2020* (Benefit and Monitoring Report Number 3).

The increase in local revenues has also created resources for funding local public services and development. UGIIP requires participating *pourashavas* to develop their own GAP, PRAP (which may include microcredit and skills training), and slum improvement plan. UGIIP goes a step further than the typical urban development project and requires participating *pourashavas* to identify in their annual budgets the funding source for the development plans and show proof of annual disbursements during the project. The funding sources must be from local revenues and not central government transfers. This strategy aims to regularize *pourashava* allocations and spending, rather than creating a dependence on limited project-based funds or the unpredictability of the central government budget transfers. See Appendix 8 for an example of the budget reporting from UGIIP-3 *pourashavas'* standing committees on women and children.

The overall annual budget preparation and disclosure process has gained new transparency and accountability under UGIIP. Estimated budgets are discussed with the town-level coordinating committee (TLCC), receive public comments, and are modified to reflect comments before being approved by the *pourashava* council. The approved budget is disclosed on the *pourashava* website. Annual income and expenditure statements and audit of accounts are done by a standing committee.

Service Capacity

The success of the reforms and technical assistance for modernizing *pourashavas'* administrative systems and training human resources is evident in the satisfaction levels of residents. According to the 2020 BME surveys, 80% of respondents had contacted the *pourashava* in the previous year. Most conducted business personally (63%); 34% conducted business through e-mail, letters, or other communication channels; and 4% by phone. Nearly 76% reported feeling either

> *"The UGIIP-3 activities have been much more visible, and monitoring has been much stronger because the money they are spending on their development is their own."*
>
> **Nilufar Yasmin**
> UGIIP-3 gender team leader

satisfied or highly satisfied with *pourashava* services and communication, and 88% said their issue or concern had been resolved. Public satisfaction is also evident in basic services.

Table 7 offers a snapshot of the scope of training topics that the project supported and volume of *pourashava* stakeholders who participated (as of 2021, with training ongoing). Trainings have involved diverse participants, methods, and topics.

Financial management. New financial management systems are increasing transparency and public accountability. *Pourashava* accounts are fully computerized and accounting reports are regularly generated, along with regular updates of the inventory of fixed assets, the rental and lease values of properties, and the depreciation fund account. Depreciation accounting and maintenance of a depreciation fund for replacing equipment as needed is not a standard practice in Bangladesh. UGIIP-3 has introduced depreciation accounting, though it does not require it. UGIIP project staff report that *pourashavas* are trained and equipped in the practice, though it is not being universally taken up by UGIIP-3 *pourashavas*. More training, advocacy, incentives, and demonstrations by champions and adopters of the practice could lead to its wider adoption.

Computer systems have reduced losses and corruption and have helped gain public confidence, reported Dewan Kamal Ahmed, mayor of Nilphamari *Pourashava* and president of the Municipal Association of Bangladesh.

That *pourashavas* are spending their own revenue on development activities is having a positive effect. Table 6 summarizes current spending levels on basic services from local revenues and their corresponding satisfaction levels, according to BME survey results. The gender team leader for UGIIP-3, Nilufar Yasmin, noted, "The UGIIP-3 activities have been much more visible, and monitoring has been much stronger because the money they are spending on their development is their own."

Table 6: UGIIP-3 *Pourashava* Spending on Basic Services and Public Satisfaction Rates

Basic Service	Total Spending from Local Revenues during UGIIP-3 Implementation	Public Satisfaction Rates	
		Pre-Project	Post-Project
Solid waste	More than $13 million	10.26%	90%
Sanitation	Nearly 2 million	8.61%	90%
Drains	About $9 million	10.59%	88%
Streetlights	More than $7 million	46.03%	93%
Mobile maintenance teams[b]	More than $4 million		90%
Road Maintenance	Not available	29.64%	90%
Total	**$35 million**		**90%**

UGIIP = Urban Governance and Infrastructure Improvement (Sector) Project.

[a] The public satisfaction rate is based on responses from the 50 town-level coordinating committee members reported in the UGIIP-3 Annual Report 2020, prepared by the project monitoring and evaluation team for the Project Management Office. The pre-project figures are an average of the socioeconomic benchmark survey results.

[b] There are no pre-project figures to report for mobile maintenance because the service did not exist before the project.

Sources: UGIIP Project Management Office. UGIIP-3 Annual Report 2020 (for terminal figures) and socioeconomic benchmark surveys (for pre-project figures).

Training. By their participation in the committees, ordinary residents have transformed local governance and have also been personally transformed through the various trainings and the implementation experience (Table 7). Trainings of the standing committee members as well as for livelihood skills have raised general capacity for budget management across *pourashavas*. It is also not uncommon for women to have financial responsibilities on committees, in addition to their new roles as small business owners. Akand said, "Before there were any resident committees, I don't think most people could have handled projects worth Tk50,000, and the women would have especially struggled with this amount of money with their experience having been more limited to the home. Many had never seen that kind of money in their life," he said. "Now, they could handle Tk5 million, or even Tk10 million. They have funds and bank accounts, and manage the books. They advocate for themselves. These are the real social changes, societal changes that have happened because of UGIIP."

Table 7: **Volume of Training**

Topics	Participant Types	Number of Trainings	Number of Batches	Number of Trainees			Number of Training Days
				Men	Women	Total	
Governance improvement (e.g., resource mobilization, financial management)	Mayor, councilors, *pourashava* officials, community leaders, members of the standing committees	26	796	17,054	7,081	24,135	27,797
Gender development and poverty reduction	Mayor, councilors, *pourashava* officials, community leaders, members of the standing committees, civil society representatives	21	366	5,546	3,603	9,149	9,817
Preparing for and implementing reform agenda	Mayor, councilors, *pourashava* officials, members of the town-level coordinating committees	9	28	1,040	150	1,190	1,406
Infrastructure management, implementation	Mayor, councilors, *pourashava* engineering and planning officials, contractors	25	123	3,270	127	3,397	5,355
Total		81	1,313	26,910	10,961	37,871	44,375

Sources: Urban Governance and Infrastructure Improvement (Sector) Project Management Office. The numbers are a compilation of results from a series of benefit monitoring and evaluation reports.

Planning. All UGIIP-3 *pourashavas* have completed master plans, and planning units have also been established in each UGIIP-3 *pourashava* to better control building construction and land development, conforming to the *pourashava* development plan (PDP) and master plan. The master plans function as a road map with guidelines for physical development within the *pourashava* while the PDPs prioritize the development projects.

Newly standardized practices include obtaining prior *pourashava* approval of any plan for the construction of a building structure; and collecting fees for approvals, permits, supervision, and violations. In addition, upon approval of the plan (layout, structural, and essential services), a specified *pourashava* official from the engineering section visits the construction site to verify adherence to the approved plan and whether construction methods conform to standard practices to prevent hazards.

The improvement in planning at the *pourashava* level has led to other governmental institutions obtaining *pourashava* endorsement for their development work and programs. Planning has proven to also be a revenue stream for *pourashavas* through the charging of fees for building plan approvals and supervision.

Operation and maintenance. *Pourashava* operation and maintenance (O&M) plans have also been prepared and approved, with an annual 5% increase in budget allocation. To modernize their O&M of basic services, *pourashavas* spent UGIIP-3 project funds on a variety of equipment such as garbage dump trucks, industrial street vacuums, and excavators.

Resident Awareness of and Participation in the Standing Committees

In UGIIP-1 and UGIIP-2, the project established standing committees as mechanisms for involving the public more directly in municipal business, including but not limited to planning the annual *pourashava* budget, managing community development plans, and implementing development activities. Project staff reported that these committees in the past often ceased to function or did not function as robustly after UGIIP-1 and UGIIP-2 concluded. Under UGIIP-3, the Local Government (*Pourashava*) Act of 2009 legally requires *pourashavas* to establish the standing committees. UGIIP-3 leverages this law to activate those committees and build their capacity for effective participation. All UGIIP-3 *pourashavas* have formed the various standing committees and are holding regular meetings, maintaining records, and communicating committee actions to the public.

There are four major standing committees: (i) Residents' Awareness and Participation committees, (ii) TLCCs, ward-level coordinating committees (WLCCs), (iii) women and children's affairs committees, and (iv) poverty reduction and slum improvement committees (SICs). Each committee is responsible for preparing and implementing its own development plans. TLCCs have an average of 50 members and the Women and Children's Affairs Committee, an average of 10 members.

Figure 5: Residents' Awareness and Participation

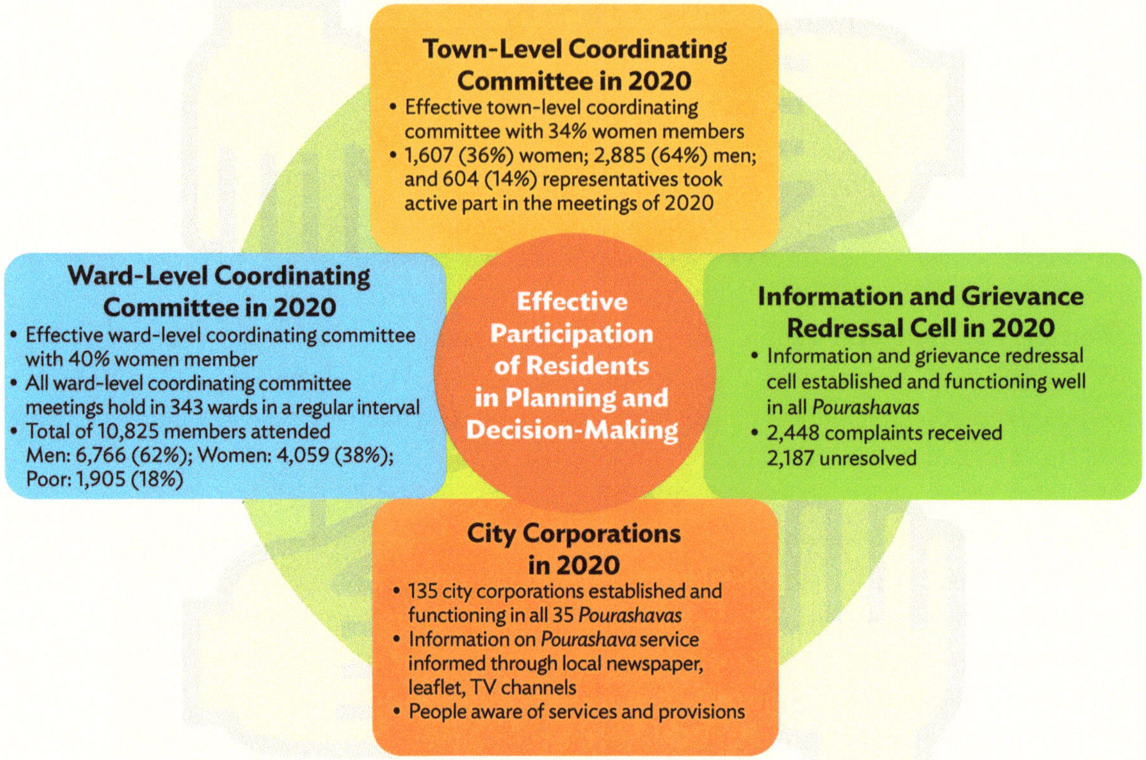

Town-Level Coordinating Committee in 2020
- Effective town-level coordinating committee with 34% women members
- 1,607 (36%) women; 2,885 (64%) men; and 604 (14%) representatives took active part in the meetings of 2020

Ward-Level Coordinating Committee in 2020
- Effective ward-level coordinating committee with 40% women member
- All ward-level coordinating committee meetings hold in 343 wards in a regular interval
- Total of 10,825 members attended Men: 6,766 (62%); Women: 4,059 (38%); Poor: 1,905 (18%)

Effective Participation of Residents in Planning and Decision-Making

Information and Grievance Redressal Cell in 2020
- Information and grievance redressal cell established and functioning well in all *Pourashavas*
- 2,448 complaints received 2,187 unresolved

City Corporations in 2020
- 135 city corporations established and functioning in all 35 *Pourashavas*
- Information on *Pourashava* service informed through local newspaper, leaflet, TV channels
- People aware of services and provisions

Source: Urban Governance and Infrastructure Improvement (Sector) Project Management Office.

The standing committees are evolving into vibrant urban forums for elected officials, community leaders, and representatives to collaborate over the issues that matter most to improving the livability of *pourashavas*. The standing committees identify and prioritize development issues, goals, and projects (along with estimated implementation costs). The respective plans of each committee are incorporated into the PDP, which is further reviewed using a participatory process to prioritize projects and finalize the budget. Among the standing committees, the TLCC has the most general scope and influence over governance.

GENDER ACTION PLAN RESULTS ▶ Gender development and empowerment has emerged as a development strategy that *pourashavas* have come to recognize and promote. All *pourashavas* have collectively spent $4.7 million on gender development during the implementation of UGIIP-3. Every year, the LGED hosts an appreciation ceremony for the champions of gender development in *pourashavas*. From 2015 to 2022, 10 of the 15 awards were given to women from UGIIP-participating *pourashavas*.

"Through UGIIP, mayors become real mayors, real leaders of the people."

Md. Shafiqul Islam Akand
former UGIIP project director (2011–2015) and former additional chief engineer for the LGED

Resident Charters and Grievance Redress

Each *pourashava* has developed a resident charter that provides public information about *pourashava* services, cost of services, and how to contact relevant municipal staff. The resident charters also remind residents of their constitutional rights and responsibilities.

By June 2021, a total of 36 resident charters had been posted in the 36 UGIIP-3 *pourashavas*, bringing greater transparency and accessibility to *pourashava* services and staff. All 36 *pourashavas* have also established protocols to encourage the practice of residents expressing their opinions on *pourashava* services and holding officials accountable.

Grievance redress cells have been formed in all 36 *pourashavas*, as a stipulation of the Local Government (*Pourashava*) Act of 2009, and records are kept of complaints and resolutions. From July 2014 to June 2021, the project *pourashavas* received more than 15,000 complaints; the 12,000 complaints (75%) that directly related to *pourashava* services were resolved.

Town-Level Coordinating Committees and Ward-Level Coordinating Committees

All 36 *pourashavas* have established TLCCs with more than 50 members in each group and 34% women's participation. WLCCs have been established in all 343 wards and, interestingly, have an even higher female participation rate of at least 40%, with 15% of members coming from low-income communities. UGIIP staff have described these committees as acting as a "mini parliament" in which residents, residents, and neighbors can bring ideas and concerns on *pourashava* activities and service delivery. During UGIIP-3, more than 70% of TLCC members attended most meetings, which are held at least once every 3 months.

The TLCCs work directly with the mayor and *pourashava* staff on gathering inputs for the PDP and annual budget, whereas the WLCCs hold regular community (or "courtyard" meetings to discuss local issues. Annual budget preparatory meetings and courtyard meetings are two major forums that bring the mayor and the public together. "Through UGIIP, mayors became real mayors, real leaders of the people. They established strong, regular connections with the people," said Akand.

In addition to their regular WLCC meetings, the UGIIP team reports that members have been active in organizing and implementing awareness-raising campaigns on coronavirus disease (COVID-19), using courtyard meetings, posters, leaflets, and door-to-door engagements. The WLCC members have also taken part in distributing food support to women and poor unemployed because of the COVID-19 pandemic.

One UGIIP staff member shared his observation that the more women have participated in UGIIP meetings, the more assertive they have become. "Female members of the TLCCs, perhaps because of religious or cultural reasons, kept their faces covered and sat on the back benches and could hardly be heard, even with a microphone," said Azahar Ali, the consultant team leader for the governance improvement and capacity building team. "After just 2 years of their participation, they grew to confidently discuss issues, participate, and share their ideas. They would tell you that before UGIIP they did not have any feelings or thoughts about their municipality. But now they say they are proud residents."

Women and Children's Affairs Committees

During UGIIP-3 implementation, the 36 *pourashavas* have allocated from their own revenue sources more than $4.7 million to support the implementation of GAPs. In total, around 12,000 low-income women from the 36 *pourashavas* have received training on different skills development, and more than 37% of them have gained full-time employment.

About 103,873 poor women and children have also received other services and support on things like basic health, schooling, and receiving bicycles to commute to school or work. "Women's issues are discussed publicly in courtyard meetings, including presentations on prevention of acid attacks, child marriage, polygamy, and family crisis management," said Azahar Ali.

Poverty Reduction and Slum Improvement Committee

Women have a role in the overall *pourashavas* as well as the SICs, which are operating in 262 slums under UGIIP-3 with 3,000 members, of which 86% of are women. They manage the implementation of slum improvement projects encompassing things such as water, sanitation, footpaths, drainage, and solar lighting. UGIIP-3 has trained almost 1,400 women and 46 men who are participating in SICs.

"SICs are the heart of women's empowerment in *pourashavas*, and women are taking charge of urban development," Akand said.

The process of SIC formation and finalization of their community action plans involves the following steps:

- identification of the slums;
- formation of the primary groups (10 to 15 per slum) comprising female members, with one in the lead role;
- formation of the SIC comprising one member from each primary group and a member-secretary purposively provided by the *pourashava*;
- formulation of a community action plan; and
- submission to respective *pourashavas*.

Aside from initial UGIIP-3 project funds, *pourashavas* have allocated $8.4 million for poverty reduction, which has come from own revenue sources to ensure sustainability.

ADB's Alexandra Vogl, the mission leader for UGIIP-3's additional financing, recalls how the women of the SICs impressed her during review missions. "They showed us hand-drawn maps of their area and had a clear idea where to put the sanitary facilities, tube wells, water pipes, and shop areas. The actually had developed a comprehensive area development plan. Also, women who participated in UGIIP-supported livelihood trainings showed us their small shops and explained how the project support changed their own and their families lives for the better," she said.

Nilufar Yasmin has worked for UGIIP for more than 15 years and on all three UGIIPs. She worked at the *pourashava* level during UGIIP-1 and UGIIP-2, before becoming the acting gender team leader for UGIIP-3. "Women are not just participating, but they are leading. They are trained, they negotiate contracts. They are in control and manage the whole thing," Yasmin said.

Pourashava-in-Focus: **Chapainawabgonj**

Impact Story 8: After Livelihood Training, "Why Not College?"

Munira Begum's story is like those of many other successful women who took advantage of livelihood trainings from her *pourashava*'s participation in the ADB-supported UGIIP. She trained, bought a sewing machine, and started earning more than she ever imagined making and selling her own clothes. Now, she trains others and loans her sewing machine to women who don't have one.

What partly motivates her now, at 37 and a mother to three children, is her past. "My parents married me off at 12. They had no means to provide for my education," Begum said. "Before turning 20, I had given birth to three children. With my husband's low income, we could hardly get food for our kids. I fell sick, malnourished, and went into depression. If I had not gotten a chance to change my fate, I would not be alive now."

Chapainawabgonj *Pourashava* has a distinguished history and is noted for its nationally famous handicrafts. Prosperity for handicraft artisans comes from the essential official government endorsement of their product, a kind of seal of approval, but this was difficult to come by for the more impoverished artisans. This changed when the *pourashava* leveraged its UGIIP participation and success to offer 6 months of training to low-income and vulnerable women. In 2016, Munira Begum, along with 30 other participants, joined the training that she says transformed her and her life.

"Women were not allowed to go outside for work, but my neighbor pointed out to me—with some hesitance—how the *pourashava* training could help women. I was determined to change my life," Begum said. "I had no other way to help us. There was a time I could not buy food for my kids. I slept outside in the road with my small children when we lost our home as a result of eviction. I wanted to secure my children's lives and educate them, so I joined the evening training."

Begum now earns Tk10,000–Tk15,000 monthly (about $115–$230) and helps other women achieve similar earnings by selling their handmade clothes and handicrafts in the local market. She was able to send all three of her children to university, as well as herself. "At the age of 37, I am attending college with my daughter. In the beginning, people laughed at me. But completing my education was my dream, and after participating in the training programs, I realized there is no age limit for education," Begum said, who expects to graduate in 2022 with her daughter.

Begum is saving for a shop in the local market to sell directly to customers, but also plans on taking orders nationally. Her vision is to connect her women's group in the Rail Bagan community to larger markets and create a sustainable business. "The women of my community are collectively trying to change their lives," she said, emphasizing that she helps herself when she helps other women. "This opportunity from the *pourashava* changed my destiny, and my success has helped hundreds of women find their success. This is my biggest inspiration. I will continue working to improve our lives."

A courtyard tradition. Women of the Rail Bagan community in Chapainawabgonj *pourashava* work on a traditional blanket. Fourth from the left is popular local trainee-turned-trainer, Munira Begum.

Never too old. After a life-changing livelihoods training, Munira Begum expects to finish college in 2022.

Source: Authors.

Infrastructure Improvements

Through a collaborative process with the various standing committees, the UGIIP-3 *pourashavas* developed publicly agreed-upon PDPs based on prioritized lists of infrastructure projects. As UGIIP is both a performance-based and demand-driven model, UGIIP-3 infrastructure funds supported priority PDP projects. Following community-based development principles, beneficiaries were involved in the planning, training, and implementation of the selected plans, activities, and projects.

Water supply and solid waste management were major investments areas of UGIIP-3 infrastructure funding, but improvements were also made in a variety of other areas, such as roads and culverts, drainage, sanitation, basic urban services to the poor, municipal facilities, and O&M. Infrastructure funding also supported new or improved municipal markets, parks, land *ghats* (boat landing stations) for water transport, bus terminals, public toilets, truck terminals, slaughterhouses, and community centers. Street lighting was also installed throughout *pourashavas*, enabling residents to move about more safely at night for work, commerce, or socializing.

In most cases, initial estimated costs of the *pourashava* development plans exceeded the expected UGIIP-3 funding allocations, requiring *pourashavas* to develop revised project proposals. In total, UGIIP-3 funded an average of 42% of the initial PDP proposals, an indication of the need *pourashavas* have for infrastructure. Appendix 7 summarizes the difference between *pourashavas'* initial and revised *pourashava* development plans.

Phased infrastructure work. The infrastructure projects were scheduled over three phases. The first phase concentrated mostly on roads and drainage. Much of the other infrastructure projects happened in the later phases. Table 8 summarizes the planned and actual improved infrastructure under UGIIP-3.[19]

Table 8: Summary of Planned and Actual Improved Infrastructure

Infrastructure Type	Planned	Actual
Improved or new urban drainage (km)	300	600
Water supply pipeline (km)	180	281
Meters (no.)	60,000	60,000
Improved roads (km)	600	1,645
Various municipal buildings (no.)	82	82 minimum
Slum upgrades (no.)	262	262
Fecal sludge management sites (no.)	14	23
Sanitary landfills (with daily capacity of 200 tons) constructed or upgraded (no.)	20	26

km = kilometer, no. = number.
Source: Asian Development Bank.

---

[19] ADB. 2014. Manila. Appendix 1: Design and Monitoring Framework. https://www.adb.org/sites/default/files/project-document/81507/39295-013-rrp.pdf.

Climate resilience sets new standard. With the support from ADB's Urban Climate Change Resilience Trust Fund, 80% of subprojects considered and, if necessary, incorporated climate-resilient materials and construction practices into their design. Roads and drainage are designed with a 10% extra provision for increased rainfall because of climate change. Other climate-responsive designs include the following:

(i) **Roads.** Road designs considered probable waterlogging stemming from climate change-induced increases in precipitation. Cross drainage structures were widened and provided at more frequent intervals to balance different water levels on either side of the road. The crest level of the roads has been designed at least 600 millimeters above the highest flood level. A substantial percentage of the roads in low areas and close to marketplaces were switched from flexible pavement to rigid pavement, which is less prone to damage from water stagnation. In designing rigid pavement, a greater number of reinforcements were used to offset temperature-induced expansion and contraction. Trees were planted alongside new roads to contribute environmental counter effects of urban carbon dioxide transmissions and increasing temperatures.

(ii) **Drainage.** The engineering designs considered making drains wider along with other hydraulic parameters to accommodate increased flow from heavier downpours caused by climate change.

(iii) **Water supply.** Production tube well bases and the location of prime mover motors, chlorinators, and other fixtures (including pumphouse floors) were elevated to well above possible inundation levels resulting from climate change. Standpipe beds were also raised above high flood levels. A 15% extra reinforcement provides the overhead tanks with additional resilience to cyclones and heavy wind.

(iv) **Landfills and public toilets.** Sanitary landfills were designed with high embankments to prevent overflow, and production tube well beds and pump houses were constructed to restrict flood waters. The superstructures of public toilets were built 0.3 meters above the highest flood level.

As a result of the climate-sensitive approaches, about 15,000 tons per year of carbon dioxide emissions will be reduced after UGIIP-3 is completed.

Roads

UGIIP infrastructure funds could be spent on (i) construction or rehabilitation of roads, junctions, footpaths, bridges, culverts, and boat landing stations; or (ii) procurement of equipment for routine maintenance, traffic management, and road safety. Road and drainage work was often paired.

Every BME survey found that the intended beneficiaries saved travel time from the construction of roads in the surveyed *pourashavas*. Pedestrians benefited the most in time savings from the new roads, followed by those travelling by rickshaw and other nonmotorized means of transportation, and motor vehicles.

New roads also contributed to an increase in commercial activities. One BME survey found that surveyed *pourashavas* had experienced a 36% increase in trade licenses and traffic along the new and rehabilitated roads. Beneficiaries also perceived an increase in their property values. Residents and commercial property owners also perceived an increase in their property values of up to nearly 60% in many cases.

Ghorashal *Pourashava*. Residents say they are proud of the town Ghorashal is becoming because of changes that began under UGIIP. Road improvements, like the one shown here, are building economic competitiveness in the town.

GENDER ACTION PLAN RESULTS ▶ Road construction was the largest employment opportunity of UGIIP-3, followed closely by drainage construction. The UGIIP-3 GAP required contractors to hire women for 20% of all jobs, but contractors have filled nearly 30% of all work positions with female laborers. Contractors also complied with GAP stipulations for separate toilets, equal pay, and other equity conditions for female workers.

Ghorashal *Pourashava*. Residents say they are proud of the town Ghorashal is becoming because of changes that began under UGIIP. Road improvements, like the one shown here, are building economic competitiveness in the town.

Drainage

Sustainable urban drainage systems are designed to reduce the potential impact of new and existing developments with respect to surface water drainage and more so to stormwater drainage. UGIIP-3 designs followed sustainable urban drainage designs for size, shape, invert, and gradient, including the extremely important well-designed outfall.

More than 80% of those surveyed in BME surveys reported that they (i) noticed a decline in the prevalence of mosquitoes, (ii) were able to keep their homes cleaner, (iii) noticed improved vehicular movement, and (iv) noticed that reduced waterlogging had improved communications. Survey respondents also reported a dissipation in bad odor from the street drains, an increase in land values, and general improvements in environmental and human health.

Pourashava-in-Focus: **Nilphamari**

Impact Story 9: After 35 Years of Working Construction, She Earns Equal Pay

Maleka Begum says she has felt a lifetime of shame on account of her being poor, abandoned by her husband, and raising children alone. Nilphamari *Pourashava*, however, has made sure she never feels shame working on its construction sites.

"It's been almost 35 years that I have been working as a construction worker," Begum said. "When we started road construction work for Nilphamari *Pourashava*, for the first time in my life I was getting equal pay. My daily wage of Tk400 is equal to that of men working on the same project."

Since 2016, Begum has been managing 20 laborers, including five other women, working on a UGIIP-supported landfill construction site. The project has introduced workers to guarantees of workplace equity, including equal pay for equal work between men and women. "When we get equal pay, we realize our contribution is equal," Begum said.

A mother at age 14 and a single mother for most of her two children's lives, Begum said she's faced all kinds of poverty, humiliation, and injustice. The equal pay and worksite conditions has given her pride, though. "I never feel shame working on the roadside or in a landfill," Begum said. "Women in my group often tell me that they feel ashamed of their work. But I tell them, our jobs are equal to those of men."

The equal pay has boosted her team's confidence, she said, proud that Nilphamari *Pourashava* is influencing other construction sites to endorse equal pay and bring gender diversity to work. Under UGIIP, construction workers have been assigned to projects building roads, a highway, a landfill, and a drainage system. Tax revenue increased 118% during 2020–2021 because of continual development in the city.

"The face of our city is changing every day. I feel happy when I see the newly constructed road and ongoing progress," Begum said. "Recently, I married off my only daughter. I can provide for my children with my savings from the site work. What can be more rewarding for a single mother?"

The returns from equal pay. Maleka Begum says her team has experienced more confidence as a result of earning pay that is equal to their male coworkers on project construction sites.

Source: Authors.

CHAPTER 3: THE LATEST RESULTS

Dry from drainage. Residents of the Hotat Para slum in Nilphamari *Pourashava* enjoy a cleaner, drier environment from UGIIP-supported facilities such as drainage, toilets, and streetlights, which also make the community safer and more mobile.

Drainage improvements were of value, especially in slum communities. Residents had been vocal during consultation about the need for sanitation coupled with drainage to reduce the risk of disease, among other benefits. Covered drains also increased accessibility in dense urban areas, allowing for vehicles to transport goods that were otherwise carried into communities previously as headloads. Drains and footpaths have also helped connect roads to increase mobility and direct access.

Dry from drainage. Residents of the Hotat Para slum in Nilphamari *Pourashava* enjoy a cleaner, drier environment from UGIIP-supported facilities such as drainage, toilets, and streetlights, which also make the community safer and more mobile.

Pourashava-in-Focus: Sherpur

Impact Story 10: Business Has Boomed with Safe, Dry Passage from New Road, Drainage

Tera Bazar Road in Sherpur *Pourashava* has a history of hindrances. Before the *pourashava* began rehabilitating the road, people could hardly walk down the street or in its alleys because of the surface condition and the garbage strewn everywhere. During the monsoon season, residents could not leave their homes and shopkeepers could not get to their stores.

Malay Chaki, a local business owner, recalled times when people fell into the open gutter. No one wanted to rent his three shops because of the 2–3 feet of waterlogging that occurred every rainy season. Floodwaters would typically stay more than a week. Rarely would one see a family outside, as the place had become a garbage field.

The value of each shop is more than Tk3 million (about $315,000), Chaki said, and no one would ever think to sell their property because of the recurrent increase in market value. Despite environmental conditions, land was holding its value. At the same time, "People were not ready to give rent of Tk1,500, even if I offered them the option to pay no security deposit," Chaki said. "We just struggled. There was no way to improve our living conditions. Since UGIIP started, people have started earning more. So paying tax is no longer a big issue for businesses here."

Sherpur *Pourashava* invested its UGIIP funds into roads, sewerage, drainage, and solid waste management (aside from dedicated funds for social programs). Since Tera Bazar Road was rebuilt in 2016, business has been booming. "Low-income people are getting more work because now shop owners are expanding their businesses," Chaki said. "In my shop, employees come from other districts, too. This place has turned into a great economic hub now." Whereas he was not able to rent his shops for Tk1,500 before the road construction, "Now every shop owner gets a monthly rate of more than Tk18,000, which is an annual increase of about 90% over 5 years," he said.

Shop owners and residents also meet monthly to discuss issues and opportunities for improving the local environment. "There was a time when people threw garbage here and there. Now we have a waste management system," Chaki said. "UGIIP-3 changed people's hygiene behavior and made us responsible residents, which I think is revolutionary."

Finally, business is booming. Malay Chaki, a local business owner on Tera Bazar Road, has been able to rent out his store space and welcome more shoppers because of improved roads and drainage.

Clean and dry. Tera Bazar Road no longer invites floodwater or garbage because of the rehabilitated road surface and drainage structures.

Source: Authors.

Pourashava-in-Focus: **Khagrachari**

Impact Story 11: Infrastructure, Livelihood Trainings Bring Together Diverse Residents

The ethnic minority community of Kaladeba is a different place after Khagrachari *Pourashava* undertook governance reforms as a means to earn infrastructure funding from the ADB-supported UGIIP.

"We now have toilets, tube wells, and streetlights. All of these happened because of UGIIP," said Rupa Marma, 40, a Kaladeba community leader. "The road has saved us from suffering on rainy days. Now we are living like human beings."

Khagrachari *Pourashava* has utilized UGIIP infrastructure funds for building a landfill and new roads, developing water sources, and providing market space that is especially welcoming to indigenous vendors. The livelihood programs have made a personal difference in the lives of many.

For 1 month, along with 20 other trainees, Marma learned mushroom farming from a UGIIP-supported program. "I decided to participate because I knew we were already a very vulnerable community. I wanted to do something and inspire other women in my area," Marma said. "And since then, my life has been transformed."

Before she attended the training, Rupa had never earned any income. She has three children, and her husband's income from working in a furniture shop has never been enough. "I had no income 5 years ago. I was unable to buy anything for my children or myself without my husband's contribution," she said. "Now I earn Tk10,000–Tk12,000 monthly (about $115–$130), which I can easily contribute to my children's education. For my whole life, this was unimaginable to me."

Discovering leadership. The project-supported livelihood program helped Rupa Marma, 40, gain courage.

continued on next page

> **Impact Story 11** *continued*

Her financial independence has helped her grow in other areas of her life. She now organizes monthly meetings for her ethnic minority community to discuss any maintenance needs on infrastructure and utility connections the community has gained under UGIIP. Every evening, women gather in Marma's yard. After seeing Marma's accomplishments, other women of Kaladeba also started producing mushrooms with her help. "I feel happy to show other women how to do mushroom farming. It is a matter of great happiness for me to become a part of this change," Marma said.

Since 2013, more than 1,965 indigenous and Bengali residents have trained together to learn various skills in areas such as sewing, driving, computers, and mushroom farming. "There was always a conflict of interest between local groups," Marma said. "But these training programs and our monthly meetings have helped us make a better bond. We are working together for our families and our community. I feel proud to be part of this change."

A place for growth. Residents of the Kaladeba ethnic minority community say they are more financially secure because of project-related livelihood programs, new infrastructure, and community solidarity.

Training together. Since 2013, more than 1,965 indigenous and Bengali residents have trained together to learn various skills in areas such as sewing, driving, computers, and mushroom farming.

Source: Authors.

Water Supply

Water supply subprojects aim at expanding the service area and improving service quality for better access to safe drinking water. Table 9 summarizes the scope of water supply subprojects implemented in 27 of the 36 project *pourashavas*.[20] O&M plans and systems were also improved. Volumetric water tariffs are being introduced in all the 27 *pourashavas* with water supply subprojects, leading to an average tariff collection efficiency of 77%.

One of six. Sherpur *Pourashava* used UGIIP-3 investment funds to build a water tank with a capacity of 680 cubic meters, increasing coverage to 73.8%. The tank was one of six built across the UGIIP-3 *pourashavas*.

Table 9: Summary of Improved Water Supply Infrastructure

Water Infrastructure Type	Actual	Notes
New water mains (km)	375	In 18 *pourashavas*
Production tube wells (no.)	86	Construction, rehabilitation, and regeneration
Hand tube wells (no.)	150	Construction, rehabilitation, and regeneration
Household meters (no.)	60,000	
Overhead tanks (no.)	7	Each with a capacity of 680 cubic meters
Water treatment plants (no.)	1	For arsenic and iron removal from the groundwater in Laxmipur *Pourashava*; capacity of 300 cubic meters per hour
Tube wells (no.)	150	

km = kilometer, no. = number.
Source: Asian Development Bank.

---

[20] Six *pourashavas* did not pursue UGIIP support for water supply subprojects for the following reasons: (i) three hill districts were excluded because of the Department of Public Health Engineering's dedicated input in those districts from the Chattogram Hill Tracts; and (iii) three towns (Benapole, Nabinagar, and Tungipara) were excluded because of being covered by other Government of Bangladesh projects. Meherpur was dropped because of a water hardness problem, Cox's Bazar was dropped because of unavailability of land, and Chatak's water supply subproject was dropped because of a delay in the design of a water treatment plant and pipe network.

Solid Waste Management and Sanitation

UGIIP-3 has improved scientific waste segregation, collection, and disposal at a scale unlike any previous effort of the LGED. Sanitary landfills have been constructed or improved in 27 *pourashavas*, along with fecal sludge treatment plants in 23 *pourashavas*. *Pourashavas* are generating compost from the collected organic solid waste from kitchens and the treated sludge from fecal sludge treatment plants. Funds have also supported the procurement of solid waste collection trucks and other maintenance equipment, the construction of 51 public toilet facilities, and public hygiene awareness campaigns.[21]

GENDER ACTION PLAN RESULTS ▶ New facilities constructed with UGIIP-3 support (as well as some existing facilities) incorporate female-friendly design features. Female beneficiaries participated in consultations, focus group discussions, and standing committees to identify the kind of infrastructure and amenities they need.

The overwhelming majority of all bus stands, train stations, and ghats (boat landing stations) now ensure the provision of separate toilets, waiting space, and ticket counters for females. Although UGIIP-3 funding supported the construction of only one park, 14 parks in the *pourashavas* were included and now have sufficient lighting for public safety, and separate toilets and sitting areas for females.

Nearly all of the *pourashava* municipal buildings now offer private workspace for female employees, and all of them offer toilets for females. UGIIP established what was originally called a "women's corner" in the municipal buildings. "But women did not like that," said Md. Shafiqul Islam Akand, former UGIIP project director (2011–2015) and former additional chief engineer for the LGED. "They didn't like the word 'corner.' They did not like how that sounded—put in a corner." So *pourashavas* adopted other names for areas that service female employees and clientele.

---

[21] By the end of 2021, 12 of the 26 sanitary landfills and 10 of the 23 fecal sludge management facilities had been completed. Extensive training and demonstration are ongoing, covering a wide spectrum of solid waste management, including generation, transportation, and disposal of solid wastes.

> **Pourashava-in-Focus: Chandpur *Pourashava***

Impact Story 12: With Access to Safe, Private Toilets, Women Move On to Other Causes

Nazma Begum, 30, grew up in Puran Bazar Refugee Colony in Chandpur *Pourashava* with about 450 families. Over the years, as the camp population grew, she felt the camp living space tighten. "The place was unhealthy, dirty, damp, and unhygienic," she said. "But now we are better able to maintain hygiene thanks to the new toilets installed by UGIIP."

Nazma Begum and other leaders, like Rabeya Begum, are actively involved in leading their community development group inspired by UGIIP. UGIIP's provision of new toilets and drainage was the first major change to their physical environment. Before the project, the community had only three toilets for 2,500 people. "We had to queue for the toilet. Girls could hardly use the ones we had," Nazma Begum said, becoming emotional recalling her experience.

Parents living in the community had to regularly take their children by boat to the hospital because of diarrhea and other waterborne diseases. The improved sanitation and the improved environment have resulted in only rare cases of related illness, and children are now able to study and play with fewer worries. "What can be more joyful than seeing their good health and growth as a parent?" Begum said.

UGIIP has also helped women organize for various livelihood and self-improvement trainings in areas such as financial literacy for small business owners. Nazma Begum is in charge of her community's monthly savings treasury. The savings group has 25 members, while other groups have as many as 50 members. "Each of us saves Tk200 monthly (about $2). After 5 years, at maturity, every group member received Tk20,000 (about $230). Some of the women from the group invested the money into their own business, such as clothing or beauty care making," Nazma Begum said. After more than 5 years, the savings group is still enabling women in leadership and community improvement. "We want a better future for women," said Rabeya Begum, another community leader. The women believe Chandpur *pourashava* is advocating for women's rights and participation on even larger scales.

Leading by example. Nazma Begum, 30, leads a local savings group in Puran Bazar Refugee Colony in Chandpur *Pourashava*.

Increasing standards. UGIIP-supported infrastructure in Puran Bazar Refugee Colony has motivated local community groups to maintain cleaner, drier, and safer communal areas.

Source: Authors.

Municipal Facilities

Aside from basic infrastructure, UGIIP funds also supported a variety of municipal structures that had the potential to earn *pourashavas* revenue or create savings. These facilities included community centers, bus and truck terminals, street lighting, kitchen markets and slaughterhouses, shopping centers (or municipal markets), ghats (boat landing stations), and recreation centers. Many *pourashavas* replaced energy-consuming incandescent streetlights with energy-efficient sodium lights or LED bulbs. Solar panel-powered lighting systems have also been installed with the project funds. Multistorey multipurpose municipal markets were found to be a strong revenue source for offsetting O&M expenses of other municipal infrastructure.

A welcome market.
Balang Tripura is one of many ethnic minority vendors who take advantage of a weekly market organized by Khagrachari *Pourashava* for distant indigenous communities to come and sell their products. The market was constructed with project funds.

An intercity hub.
The UGIIP-supported Ostomitola Intercity Bus Terminal in Sherpur *Pourashava*, which is under construction, is the only bus terminal in the city that will have the capacity for 60 buses at a time.

Pourashava-in-Focus: **Sherpur**

Impact Story 13: Club Promotes Female Athletics, Entrepreneurship, Leadership

It was only a few years ago that Tasfia Tabbasum Oyshee found riding a bicycle or wearing a sports jersey challenging, she said. Now, she is a national badminton champion. "Our *pourashava* actively promotes girls and teenagers, and we are finding a way to dream," she said.

In 2017, Sherpur *Pourashava* leveraged some of its funding from the ADB-supported UGIIP to develop a creative and economic space for girls and women to excel at athletics, entrepreneurship, and civic leadership.

The Pauro Kishori (Civic Ladies) Club offers training programs and various material support and networking fairs to connect female entrepreneurs to markets and promote female-led businesses. "We are united to support women facing adversity in our *pourashava*," Shabina Zaman, one of the club managers, said. "We regularly arrange programs, fairs, and other activities for promotional activities, and this is connecting buyers and sellers."

In 2020–2021 alone, Sherpur *Pourashava* trained 314 girls and women from low-income households in sewing; salon services; computers; and sports such as badminton, soccer, cricket, tennis, billiards, and table tennis. After 1 year of training, Oyshee earned a gold medal in a national tournament in school badminton. "Besides winning various games or participating in cultural programs, we are also involved in social causes," Oyshee said. For example, she and other club members campaigned to end child marriages and stopped two child marriages by involving the *pourashava* government.

Members of the club also achieved recognition in local and national dance and musical competitions. The *pourashava* sponsors local competitors' accommodation and transportation expenses during training programs and cultural events.

The Pauro Kishori Club has 66 members in addition to general participants and 35 volunteers. "Our club is diverse," said Asha Muni, the club president. "Girls from different socioeconomic backgrounds and conditions join together and nurture the talent we have."

A space for young women rising. Sherpur *Pourashava* used its UGIIP funding to provide a space for supporting young women through training and affinity groups.

continued on next page

Impact Story 13 *continued*

National champion. After only 1 year of training with the *pourashava*-supported Pauro Kishori Club, Tasfia Tabassum Oyshee earned a gold medal in a national tournament in school badminton.

Club president. Asha Muni is the president of the Pauro Kishori Club, a *pourashava* program promoting female empowerment through sports, livelihood, and leadership training.

Source: Authors.

Basic Services to the Urban Poor in Slums

UGIIP-3 funded almost $22.65 million in improved infrastructure in 262 slums. "The proportion of UGIIP funds invested into slums was relatively little, yet they had a very high impact," Akand said. The poverty reduction and slum improvement committees (SICs) in each *pourashava* identified priority projects, which were included in the PDP and allocated specific municipal funds. The SICs contributed in-kind labor to some projects and are responsible for construction and O&M of infrastructure. The ward-level coordinating committees, SICs, and other local committees meet monthly on average to address local environment and infrastructure issues. Slum improvement infrastructure has included the construction or installation of footpaths, drains, water supplies, street lighting, community latrines, and waste bins (Table 10).

Table 10: Slum Improvements

Improvements	Achievements
Hand pumps (no.)	1,197
Solar streetlights (no.)	647,647
Area lighting (no.)	820
Toilets (no.)	2,942
Footpath (kilometers)	102
Drains (kilometers)	51.85
Dustbins (no.)	266
Stairs (kilometers)	2.33
Tree planting (no.)	12,870
Low-cost housing units (no.)	52

no. = number.
Source: Project Management Office for the Third Urban Governance and Infrastructure Improvement (Sector) Project.

A school of their own. Low-income students have a closer school with the UGIIP-supported construction in 2010 of the Amir Uddin Pauro Primary School in Bhairab *Pourashava*, where 300 students attend.

Pourashava-in-Focus: **Moulvibazar**

Impact Story 14: She Sleeps Without Worry About Floods Washing Away Everything

After getting married 11 years ago, Suhena Akter moved to the banks of the Manu River, where her husband's family and 100 other households depended on fishing. Then one night, a severe flood washed everything but their lives away.

If income from fishing was hard enough before the flood, it became impossible after. Akter's husband became ill after losing their house in the flood. "Days were tough. I starved with our children. There was nothing I could do, as women of my area had no means to earn an income," she said.

In 2017, because of Moulvibazar *Pourashava's* participation in the ADB-supported UGIIP, life began to change for Akter, starting with attending sewing classes and *pourashava* committee meetings that were developing plans for the town.

As part of Moulvibazar's GAP developed under UGIIP, the *pourashava* committee allocated funds for 39 highly vulnerable women to develop or reopen their shops. Akter received Tk10,000 initially (about $115) to start her business and open her shop. She estimates she earns Tk500–Tk1,000 taka daily (about $6–$12), enough to send three of her children to school. "We are expanding our business now, having learned financing skills from the training," Akter said while quickly serving tea and biscuits to people arriving at her shop.

Though Akter never attended school, she intends for her children to finish school. "Life was terrible. Now we have hope," she said. "We can look forward to continuing living a better life. Our town has changed, too." The *pourashava* constructed Kusum Bag Nodir Par Road followed by a bridge to protect roadside infrastructure and households. The investments have created economic security and peace of mind for women like Akter. "The river took everything from us. We are safe now only after the road was built," Akter said. "For the first time in my life, I can go to sleep without the fear of erosion or flood."

Restoring her shop. Suhena Akter reopened her store after it was destroyed by river flooding.

Expanding her services. Suhena Akter diversified her income sources by attending a project-sponsored sewing workshop. She makes and sells children's clothes to augment her income from her convenience store.

Source: Authors.

Rebuilding what the river took. Suhena Akter woke one night to the river near her fishing community washing away what little her family owned, including their house. She rebuilt her shop and diversified her income streams with UGIIP-initiated programs. Read her story on the preceding pages.

CHAPTER 4: SUCCESSES, CHALLENGES, LESSONS

The positive results and impacts from the first Urban Governance and Infrastructure Improvement (Sector) Project (UGIIP) have been confirmed by UGIIP-2 results and reconfirmed as UGIIP-3 nears completion. As the Asian Development Bank (ADB) and the Government of Bangladesh consider future investments in municipal governance and infrastructure support, this chapter summarizes the success factors, challenges, and lessons from implementing UGIIP.

The UGIIP approach has aged well over its 20-plus years of implementation, with continuous improvements over the years. All three UGIIPs have proven what works, what does not, and what still needs figuring out. The success factors discussed in the following chapter should give project designers and governments enough courage and reason to raise their expectations of what local governments can accomplish regardless of their baseline resources and capacity. UGIIP has proven that gaps in both resources and capacity can be overcome quickly with equal parts support, willingness, and public accountability.

There have been perennial challenges that seem to have been resolved only nominally and incrementally, such as issues related to centralized control of human resources, recruitment and retention of qualified staff in *pourashavas*, a qualified national consultant pool, and the prioritization of operation and maintenance (O&M). More ideas and innovations are still needed. The UGIIP-3 *pourashavas* were a particular challenge because they were smaller and less developed. Most of the *pourashavas* that have not yet participated in UGIIP are lower-class municipalities and will require more capacity development and technical support.

> *"I experienced an extraordinary collaboration. Because of the long history between ADB and the LGED, it was a trusted partnership."*
>
> **Alexandra Vogl**
> UGIIP-3 mission leader for additional financing and current principal planning and policy specialist in the Strategy, Policy, and Business Process Division

The greatest success factor in UGIIP, and common to most projects, has been the capacity of the executing and implementing agencies. All former ADB UGIIP mission leaders and project staff interviewed for this report identified the Local Government Engineering Department (LGED) as the single greatest success factor. "I can say that no matter what the ADB project is, the most important success factor is our executing agency," said Masayuki Tachiiri, UGIIP-2 mission leader and current director of the Strategy, Policy, and Business Process Division of ADB's Strategy, Policy, and Partnerships Department. "We can do good design and due diligence, but if we have a less capable executing agency or counterpart, we can't accomplish anything. Our project is going to go nowhere. We were very fortunate with this project to have a very capable, very reasonable, clean and transparent government agency."

Of processing additional financing for UGIIP-3, Alexandra Vogl, the UGIIP-3 mission leader for additional financing and current principal planning and policy specialist in ADB's Strategy, Policy, and Business Process Division, said, "I experienced an extraordinary collaboration. Because of the long history between ADB and the LGED, it was a trusted partnership. Of course, there were discussions and convincing on both sides of what we should and should not do, but there was also a common understanding of the kinds of impacts and development results we wanted to achieve."

Park in Tungipara. A riverside park constructed in *Tungipara pourashava* has a riverside walkway and an amphitheater for open-air programs

Pourashava-in-Focus: **Maulvibazar**

Impact Story 15: Once Timid to Touch a Computer Key, She Is Training Hundreds of Others

Rokasana Nasnin was once hesitant to touch the button of a computer. Now, she trains hundreds of other young people from low-income households in basic computer skills.

This is a huge opportunity for students who have no means to invest in learning computers," Nasnin said. "Without computer knowledge, the younger generation has no future. One-hundred percent of the participants arriving in our lab come from extremely poor households."

In 2019, Nasnin and 30 other low-income community members enrolled in the computer training program offered by Maulvibazar *Pourashava* under the ADB-supported UGIIP.

Nasnin has been working as a *pourashava* trainer for almost 1 year. She teaches six batches of 30 students in each batch. There are as many as 20 batches in a year. Female participants feel more enthusiastic and confident about their performance when they see Nasnin will be their teacher.

"We always get 50:50 girls and boys attending our lab," Nasnin said. "This is huge because getting female participants from lower-income households is a remarkable change."

Mohammad Mustakim Ahmed, 22, also completed 3 months of computer training at the end of 2018 along with the other 22 participants. "That training program helped me find my vision. After completing the basic course, I enrolled in a web design course. Now I am working with an agency that has international clientele," Ahmed said.

Computer role model. Young women from low-income communities are enthusiastic to learn computer skills from their female trainer, Rokasana Nasnin. Like them, Nasnin was once afraid to even touch a computer. The computer lab and training program are outcomes of a UGIIP-supported initiative of Maulvibazar *Pourashava*.

continued on next page

> **Impact Story 15** *continued*

Ahmed now earns an average of Tk30,000–Tk50,000 monthly (about $350–$500), depending on the work he gets through his agency. Ahmed wants to invest his savings in the information technology sector and wishes to have an agency that will create more chances for lab participants in the future. "If I did not get chance to take computer training, I would never dream to do what I am doing now. Life would be very different," Ahmed said. "Everyone deserves this chance regardless of their economic background."

Computing greater chances. Residents of the Kaladeba ethnic minority community say they are more financially secure because of project-related livelihood programs, new infrastructure, and community solidarity.

Everyone deserves this chance. After completing a 3-month computer course offered as a UGIIP-supported *pourashava* program, Mohammad Mustakim Ahmed enrolled in more training and found regular work for a web design agency. He trains others, believing that "everyone deserves this chance."

Source: Authors.

1. What mayors initially feared has been their greatest success: public accountability

A major challenge in the beginning of UGIIP-1 was the confidence of municipalities in their ability to pay their bills in full, repay debt, find new local revenue sources, improve their tax collections, or involve the public in official municipal business.

"They didn't think they could do it," said Md. Shafiqul Islam Akand, former UGIIP project director (2011–2015) and former additional chief engineer for the LGED. "Municipal leaders asked 'Why? Why involve the public and civil society?' They had a negative mindset. They were afraid of criticism. But after 2 or 3 years, these same mayors realized how much they were benefiting from UGIIP."

The consultative and participatory mechanism of the town-level coordinating committees (TLCCs) and ward-level coordinating committees (WLCCs), which include elected officials and representative resident groups, was the driving force for *pourashava* officials to meet the required criteria. Their close involvement played a major supporting role in disseminating information about the reforms, the obligations, and the benefits of local infrastructure improvements (if the *pourashava* met the governance criteria). Governance and infrastructure improvements were, therefore, achieved because of a collective effort. The TLCCs and WLCCs kept officials accountable to their project agreements. Akand summed up the UGIIP approach as follows: "motivate, convert, make an example out of them."

When Tachiiri began processing UGIIP-2, he was surprised by what he found. "I thought this was not a popular project among mayors because it is a lot of work, but that was not true," he said. "Many mayors liked this project. They used UGIIP as a good platform to do the work, and they gained the support of the residents. We know that because we used resident report cards."

2. Designing for results and momentum offers mayors a chance at good results within their political tenure

Inherent in good governance are political benefits. Mayors were most often able to demonstrate improvements within their tenure because of participating meaningfully in the project. In sharing the decision-making processes of the *pourashavas*, mayors were working alongside—often in the same room as—many different kinds of residents and resident groups. The process built multidimensional community relations while increasing each mayor's visibility and credibility with the public.

3. Vigilance against political interference in the evaluation of *pourashavas'* performance protects the credibility of the entire program

Ensuring that a *pourashava's* reforms were authentic and effective meant following a strict evaluation of their performance against the criteria and indicators. Project staff reported that safeguarding the process from politics was challenging, but imperative. The project director acted as gatekeeper to an independent, fair, and impartial evaluation.

> *"Before this project, the mayors and the engineers in the pourashavas had limited exposure to how a municipality could support gender development…"*
>
> **Norio Saito**
> UGIIP-3 mission leader and current director of ADB's South Asia Urban and Water Division

4. Performance-based access to funding can create healthy competition

UGIIP-1 set the precedent for both success and failure in a high-stakes system. Succeed in meeting the incremental improvements in basic good governance, and a *pourashava* would qualify for its allotment of infrastructure funding. Fail, and the *pourashava* was disqualified, allowing other *pourashavas* to compete for its infrastructure allotment. UGIIP-1 administrators and evaluators proved the rule's seriousness when six *pourashavas* were disqualified from the project for failing to pay what they owed to public utilities. Two of these *pourashavas* later rejoined the program in UGIIP-2 and no other *pourashava* has been disqualified since, though some have come close. UGIIP-2 and UGIIP-3 allocated more support for local capacity development and technical assistance to support *pourashava* success. When a *pourashava* was at risk of disqualification, they were given more time and support. They were not allowed to progress to the next phase (e.g., receive infrastructure funding) until they earned a satisfactory score on their governance evaluations.

5. Intensive project support at the local level is the difference between ceremonial and effective governance

ADB has learned from UGIIP-1 and UGIIP-2 that project *pourashavas* need local, high-quality, front-loaded capacity building and strong institutional support. UGIIP-3 began capacity building of project *pourashavas* during the project preparation phase and has continued it throughout implementation. "We—ADB, government, and consultant implementers—are not administering a program. We are working together, rigorously, with *pourashavas*," Akand said.

As the UGIIP strategy has become more well-known and popular, mayors of non-project *pourashavas* have taken their own initiative to implement the UGIIP governance improvements, especially once the Local Government (*Pourashava*) Act was passed in 2009 that mandated the formation of the standing committees. The team implementing UGIIP has observed that the *pourashavas* trying to emulate the UGIIP reforms without the support of the project have generally not been able to implement the reforms as effectively. Many times, the standing committees are established, but do not function as effectively as those established by *pourashavas* under the project. The non-project *pourashavas* have also needed support in identifying opportunities for raising local revenues and executing their implementation, as well as improving their tax collection rates. They are also operating without the new administrative systems and support for computerized accounting and tax collection that the project *pourashavas* have access to.

6. A visible municipal presence and oversight of contractors is the best quality assurance

In UGIIP-3, the Department of Public Health Engineering and LGED conducted technical trainings for *pourashava* staff, field-level engineers, and municipal engineer consultants working on water supply subproject implementation and quality assurance. During a review mission in April 2021, ADB identified two effective approaches to on-the-job training and quality assurance. First, the Department of Public Health Engineering Project Management Office's project manager and executive engineer were regularly present in the field to observe implementation and note where training and capacity development was further needed. Second, in addition to training *pourashavas* to take over the O&M when construction was finished, the Department of Public Health Engineering field engineers were been involved in data collection and closely monitoring and supervising construction quality. An O&M training manual on water supply subprojects was prepared in Bengali for relevant *pourashava* staff.

7. Small contractors need fair competition

Despite having almost 500 construction packages, UGIIP-2 did not require a single day of extension. Project managers credit their monthly tripartite meetings between the project team, contractors, and consultants, and also the custom trainings for mayors, councilors, and officials to build their capacity for procurement and project management. Narayanganj Mayor Selina Hayat Ivy wants the procurement systems and policies followed for UGIIP to become standard practice for the government. "The tendering process under UGIIP was good. We followed the public procurement rules, but outside of UGIIP it is a monopoly system, not very competitive," she said. "We need to work on this. Small contractors cannot compete."

8. Budget allocations are proof of *pourashava* intent and actual action

UGIIP introduced to *pourashavas* several key development plans that need to be regularly updated and sustained by municipal resources as a part of ongoing *pourashava* development and good governance. The project supported *pourashavas* in developing their first *pourashava* development plan (PDP), which had to include a GAP, a PRAP, and a slum improvement plan. *Pourashavas* were also required to allocate resources from their municipal budgets for the implementation of those plans. UGIIP-3 introduced the new criteria of requiring *pourashavas* to increase their budgets for these plans by 5% over the previous year. Of particular success, during UGIIP-2, poverty reduction and slum improvement committees (better known just as SICs) seeded their microcredit organizations with their own funds. This allowed them to determine their own interest rates, which were often lower than nongovernment organizations' rates that are known to be as high as 15%–20%.

9. Mitigate disruptions to implementation schedules and regain momentum with strategic teamwork

Seasons, elections, and a pandemic were three major disruptors to the project's momentum in *pourashavas*. Two of these were predictable and need precision scheduling and contingencies to mitigate their limitations and potential for impact. In Bangladesh, much of the infrastructure work must be started and completed within one seasonal window before the long monsoon season begins, complicating construction work. The coronavirus disease (COVID-19) pandemic began in 2020 during Bangladesh's most productive construction season, affecting what would become more than a year of worksite delays and more stringent health and safety protocols.

10. Demand and dependence can be addressed by competition and cost sharing

A common recommendation from project staff reflecting on their experiences and observations implementing UGIIP has been to increase project resources to allow more *pourashavas* to participate or increase resources for more comprehensive investments per *pourashava*. *Pourashavas* will always have more funding needs than what the central government, donors, or projects can allocate. A national program, with donor support, could give more equal access to *pourashavas*, create healthy competition for funds, and sustain the governance work. Competition also induces the momentum that reforms need. Competition can inspire *pourashavas* to optimize local opportunities for revenue generation and more participatory development. Tying performance to infrastructure cost sharing is another incentive—more parity between governance improvements and infrastructure funds. Competition and cost sharing can support effective decentralization, removing lingering dependencies and the divide between resources and responsibilities.

11. Operation and maintenance requires special emphasis

UGIIP designs have increasingly put more pressure on project *pourashavas* to properly resource and implement O&M of both existing assets and new assets created from UGIIP. UGIIP-1 stated that *pourashavas* would be responsible for O&M of subproject investments. UGIIP-2 required *pourashavas* to prepare O&M manuals. UGIIP-3 has seen O&M manuals prepared in Bengali to increase understanding (project *pourashavas* that participated in previous UGIIPs updated and translated existing O&M manuals). These manuals were required (i) to describe the O&M procedures for all municipal infrastructure, services, and facilities; and (ii) to be designed for regular use by *pourashava* engineering staff. The project management office consultants have provided specific training to all relevant stakeholders.

The incremental O&M costs during implementation have not been included in the projects' costs, but UGIIP-1 established the criteria requiring project *pourashavas* to set aside a minimum 15% of the annual development budget (excluding UGIIP funding) to cover O&M expenditures each year. UGIIP-2 went a step further and required an annual O&M plan with a required budget that had to be approved as part of each *pourashava's* development plan. To achieve a "fully satisfactory" rating, which qualified a *pourashava* for the higher allotment amount, *pourashavas* had to also approve a 5% annual increase in the O&M budget.

UGIIP-3 pressed further, adding a seventh governance indicator to the Urban Governance Improvement Action Program (UGIAP) dedicated to "keeping essential *pourashava* services functional." This required action plans (entry phase), budget allocation (intermediate phase), and the O&M plan implemented to the TLCC's satisfaction (advanced phase). The fully funded O&M plans had to be funded and satisfactorily implemented for the regular collection and safe disposal of solid waste, the cleaning and maintenance of drains, street lighting and other infrastructure, and the establishment and operation of a mobile maintenance team.

The project staff report that implementation is not yet satisfactory, but the investments in solid waste management and sanitation O&M are a historical first for *pourashavas*. UGIIP-3 project staff report that only a few have been able to satisfactorily implement all of the O&M plans, mainly because of challenges with contractors and land acquisition.

Promoting and pressing for proper O&M is likely to continue to be one of the more significant challenges to urban governance work.

12. *Pourashavas* and projects need more responsive human resource protocols and a robust consultant marketplace to recruit and retain talent

Not unlike the challenges that other ADB projects have documented, UGIIP has experienced frequent turnover of national consultants. Positions as critical as team leaders have taken the project office more than 8 months to fill. Gender specialists are notably difficult to identify and successfully recruit. A forthcoming knowledge product about the ADB-supported Khulna Water Supply Project identifies the hiring of both government and project staff as an unnecessarily lengthy bureaucratic process that often ends where the process began: with an empty position. By the time the local government or project implementing agency gets approval to hire a particular staff member or consultant, the candidate is no longer available. The central government still controls how *pourashavas* organize their human resources (*pourashavas* need central government approval of organograms), what human resources they need (the number of positions and in what departments or divisions), how much they are paid, and who hires them.

Another challenge is the lack of a robust consultant marketplace, which makes competition among projects for consultants high and geographically driven. ADB's urban investments in Rajasthan, India, divided its focus between institutional and infrastructure building to build a "development ecosystem" for its vision for long-term urban development in the state. The Rajasthan Urban Infrastructure Development Project, which began around the same time as UGIIP and has also spanned more than 20 years over multiple phases, created and invested heavily in institutional development to become an efficient, nationally renowned urban financing and project management institution for the state. The Rajasthan Urban Infrastructure Development Project built a reputation for excellent project management, both recruiting senior consultants and training junior consultants, many of them engineers. This formed a leadership and technical pipeline—something that Bangladesh needs for a national-scale program to provide long-term technical assistance in governance and municipal infrastructure development.

13. Needed: town planners

This governance factor was added in UGIIP-2, yet only four or five *pourashavas* have a town planner. The master plans are key for systematic physical, spatial planning. "Town planners are desperately needed, but the position is long-term, in one place, so there is no opportunity for career progression," said Azahar Ali, the consultant team leader for the governance improvement and capacity building team.

14. Introduction of volumetric meters and tariffs is a tough challenge

The biggest challenge of the UGIIP-3 implementation has reportedly been the introduction of a volumetric water tariff to increase the water revenue collection and reduce the percentage of nonrevenue water. Cost recovery rates of water services are low, while demand for more reliable services is high and growing rapidly. A case study revealed that despite significant income constraints, households were willing to pay substantially more for improved water supply services over and above their current water bill. Women and households living in the poorest urban areas with the lowest service levels valued the improvement of water quality the most. The experienced resistance to volumetric meters, new tariffs, and billing can be mitigated through proactive communication strategies to bring users values into alignment with new changes in services.

15. Slum communities need greater equity in development spending

UGIIP has required *pourashavas* to allocate at least 5% of infrastructure spending from UGIIP on slum improvement. Project managers and mayors have felt the pressure to balance limited funds between much-needed infrastructure with general population benefits (such as drainage) and greater or more equitable spending on slum improvement (since 30% of residents in most *pourashavas* live in slums). General urban infrastructure improvements may benefit everyone in the city, though not always equitably. For example, slum community residents may benefit from improved water supply through communal taps, whereas more developed neighborhoods may benefit from improved water supply through household connections.

The inequity is often caused by the spatial limitations to engineer equitable benefits rather than the poor's willingness or ability to pay. Other improvements in slum communities, such as affordable housing, could make modern household services more possible, along with other development benefits, such as green spaces and community-based facilities and services for early childhood education, improved health services, and livelihood training. The SICs, first established by UGIIP and now by the Local Government (*Pourashava*) Act of 2009, are the mechanism for advocating, planning, and implementing more equitable development.

16. Recognize and promote women's role and contribution to development

UGIIP's contribution to increasing the visibility of women's role in local development is unquestioned and one of the first benefits that project stakeholders speak about when asked about UGIIP's impact. Having significantly raised the baseline on women's participation and leadership in community committees, project stakeholders say that municipalities need a leadership process—a leadership pipeline in municipal government—to promote women in professional positions. Female municipal employees have expressed appreciation for the added private spaces and facilities for women in municipal buildings, though they have objected on the public record to those spaces being labeled "women's corners."

Project managers and consultants say there are more opportunities to create or change policies to promote women and the poor. For example, the municipal budget form does not have a specific exclusive heading or section for gender or poverty appropriations. These expenses are part of the *pourashava* development plan, whereas specific line items in the budget, each with a mandatory percentage of budget allocation, would bring greater transparency to the budget as well as assured funds for these important elements of social development.

17. Standing committees are vulnerable to integrity gaps

The integrity of the standing committees is vulnerable to three structural flaws. First, the mayor's position on every committee may limit the autonomy of their discussions, recommendations, and decisions. Second, the mayor's responsibility of personally selecting and appointing members to the committees lacks transparency and creates potential concern for bias. For example, 33 % of TLCC members are women, but how the mayor selects them and how many of them are from poor communities is not public knowledge. Third, the lack of an audit and reporting system on *pourashava* standing committees means that committees may be more form than function. For example, Azahar Ali, the UGIIP-3 governance improvement and capacity building team leader, said he has observed that non-UGIIP *pourashavas* have established their standing committees to fulfill a legal obligation, but they are not operating as effectively as UGIIP-participating *pouashavas,* pointing to a *pourashava's* inexperience with participatory development and their need for the kind of technical assistance UGIIP has offered. An audit and reporting system would help identify where capacity building for standing committees is most needed, and would measure progress in participation and accountability.

18. Decentralize municipal staffing decisions

Municipalities have inherent limited institutional capacities because the central government limits the number of staff they can employ. Technical positions are hired and assigned at the central level, not the local level, making recruitment and retainment an issue. "When we started UGIIP, more than 50% of government positions were vacant. A loan covenant of UGIIP required that positions be filled," said Azahar Ali. "The central government transferred central staff for the duration of the project and then recalled them once the project was completed. The municipalities were back to the same problem again. After COVID-19, we now know how badly the recruitment situation is, especially in health care positions." Most municipalities also only have 2–3 medical officers on staff. This is a bureaucratic process challenge that can be overcome.

19. Leverage the law on standing committees

Municipalities should coordinate between development partners that are investing in them to utilize municipalities' standing committees and not create new ones. This lack of coordination leads to a misuse of resources and duplication. This is a missed opportunity to strengthen the standing committees. Until the development partners become better coordinated, the central government has no incentive to change what they have traditionally done.

20. Systematic knowledge sharing, and transfer of good practices

Pourashavas need a mechanism and system for regular knowledge sharing between themselves about governance issues, ideas, initiatives, and results. This knowledge product is proof of the good practices and lessons that UGIIP has to offer to urban development leaders and professionals. Regional delegations of mayors from India, Nepal, and Sri Lanka have also toured UGIIP-participating cities to learn about the governance-driven incentive approach for infrastructure improvements. The UGIIP team has hosted other government representatives and has attended conferences to share the UGIIP approach and value of community development planning.

A knowledge sharing system would begin with a strategy and include specific ways of sharing knowledge with *pourashavas* that have not had the advantage of participating in UGIIP. A knowledge sharing strategy would consider training on the effective and meaningful implementation of the Local Government (*Pourashava*) Act of 2009 as well as the principles, practices, and lessons of implementing the seven UGIAP governance criteria and their incremental indicators. The Municipal Association of Bangladesh has been an enthusiastic supporter of the UGIIP approach, gathering information on UGIIP results and impacts from its members, and sponsoring training and knowledge sharing events around the country (Box 3).

Box 3: Recommendations from the Municipal Association of Bangladesh

The Municipal Association of Bangladesh has been an enthusiastic supporter of the Urban Governance and Infrastructure Improvement (Sector) Project (UGIIP). The association's president and Nilphamari mayor Dewan Kamal Ahmed said the organization had taken up documenting UGIIP good practices among its members and sharing them with members who were not from UGIIP municipalities. As UGIIP-3 implementation nears completion, Ahmed shared the following recommendations for maintaining momentum on the gains made by UGIIP-3:

- Scale-up the UGIIP model to include class B and C *pourashavas*, where services and infrastructure are most constrained by local governance, technical, and financial capacity.
- Continue compiling and disseminating UGIIP best practices to all municipalities through workshops, seminars, meetings, and exchange visits between UGIIP and non-UGIIP municipalities.
- Continue policy dialogues to strengthen the Local Government (*Pourashava*) Act 2009 based on UGIIP best practices.
- Development partners should regularly engage in dialogue with each other and with governments to share good practices and lessons on urban governance that can be incorporated into project designs.
- Advocate with development partners and governments to allocate more resources for low-income communities from local and project budgets.

Source: Authors.

Pourashava-in-Focus: **Moulvibazar**

Impact Story 16: Her Life Was "Once Unimaginable," Says Widowed Mother of Four

"Once, it was unimaginable that women could have their own business," Rashida Begum, 40, said. A month of training to learn how to sew, make quality handicrafts, and run a small business is all it took for Begum and other women in her community to become their own agents of transformational change.

"I was married off at a very early age," she said, noting her husband died about 14 years ago. "With four small children, life was devastating as a widow. I wanted to do something to earn. Then I learned about the *pourashava's* training program."

Begum lives in the low-income community of Dhengar in Ghorashal *Pourashava* in central Bangladesh, where women have limited means of financial advancement, she said.

Begum and 20 others trained together and received a sewing machine once they finished the training. From her one-room shanty, she began taking sewing orders for just Tk70 (less than $1) to gain customers. Slowly, she extended her clientele. "My monthly income grew to Tk10,000 (about $115) from zero. I sent four of my children to schools and colleges. Now I have one university-going daughter," she said.

Begum said all of the women she trained with now have their own small businesses. She is also training other women, including her daughters. As women witness other women succeeding at business, they, too, feel empowered to also try their skills at earning money. "I feel good when I see other women are doing something," Begum said. "So many vulnerable women have had no means to earn, but this is a way of surviving for many women like me."

With her earnings, Begum has begun building a brick house on her husband's ancestral land. She plans on opening a shop, too. "Our place has changed a lot, too. We have new roads, and our town is thriving with many economic activities. The *pourashava* has been working to build this town as a model town," she said. "I feel proud to be part of this change."

Once facing economic despair. As a young, poor, widowed mother, Rashida Begum, 40, joined the project's livelihood program. She has since put her children through university education, is building a house, and is training other women.

Source: Authors.

Natural leader. Rupa Marma, 40, is a natural leader in her ethnic minority community of Kaladeba in Khagrachari *Pourashava*. Before attending a UGIIP-supported training, Marma said she had never earned any income. She is now a mushroom farmer and organizes monthly meetings for her ethnic minority community to discuss any maintenance needs relating to infrastructure and utility connections the community has gained under UGIIP. Read her story on page 56.

CHAPTER 5:
UNFINISHED BUSINESS

The popularity of and high interest in the Urban Governance and Infrastructure Improvement (Sector) Project (UGIIP) as an effective lever for greater local governance and infrastructure development are evident in the growth of the project since its launch in 2002. This growth has been accompanied by closer scrutiny and improvements in the governance indicators to progressively measure genuine progress.

As UGIIP-3 approaches its closing in 2022, the Asian Development Bank (ADB) and the government are determining how best to serve the governance and infrastructure needs of people living in pourashavas, especially those that have not yet participated in UGIIP. One thing is for sure, however: each UGIIP has scaled-up its coverage along with its expectations of government.

Project managers and mayors have hoped that future UGIIP or UGIIP-inspired programs or projects would increase the funding per *pourashava* and increase the number of *pourashavas* eligible to participate. Others believe the country has enough experience with incentive-based reforms and infrastructure funding to lead a national program supported by development partners.

A Government-Led National Governance Program

UGIIP has had great coverage, but more *pourashavas* need their turn to participate. At the same time, *pourashavas* need more access to finance for infrastructure projects to make a larger impact. Many of the projects supported with UGIIP infrastructure funds have been either stretched thinly across several subprojects or concentrated on a few priority projects.

UGIIP has demonstrated to the government how it could leverage its national budget transfers to incentivize reforms and reward *pourashavas* according to their individual performance. Mainstreaming the UGIIP strategy could assist in reducing competition among *pourashavas* by supporting a national program with additional infrastructure financing and technical support. A national program requires development partner support (i) to replicate the UGIIP strategy in *pourashavas* that have not yet had the opportunity to participate and (ii) to continue supporting those *pourashavas* that have done well with UGIIP, but require further monitoring, capacity building, and infrastructure finance.

"UGIIP is replicable within Bangladesh because the *pourashavas* are similar enough to one another—they are a similar size, have similar issues, and can 'compete' with each other," said Masayuki Tachiiri, the mission leader for UGIIP-2 and current director of the Strategy, Policy, and Business Process Division of ADB's Strategy, Policy, and Partnerships Department. The competition he speaks of is a healthy one, which can be driven and captured by performance benchmarking systems that bring an additional dimension of external accountability and transparency to how *pourashavas* or individual utilities and government agencies are performing against best practice standards and their counterparts.

Time for a National Urban Policy

UGIIP-1 supported the government in developing a draft urban policy, though it was never approved. Further advocacy for the policy in UGIIP-2 and UGIIP-3 did not achieve any progress. UGIIP's governance and capacity building team leader Azahar Ali, an observer of policy development in Bangladesh for many years, said, "There are many inherent problems that only joint development partnerships can help resolve, such as getting the urban policy approved, municipal organograms approved, and the human resource processes streamlined. General problems are still there in the *pourashavas* because policy is not changing, and it must change for there to be real sustainable development of the urban sector."

Norio Saito, UGIIP-3 mission leader and director of ADB's South Asia Urban and Water Division, believes that a national urban policy also needs to be accompanied by other policy reforms, particularly with *pourashavas'* own revenue collection and human resources management systems. "We need to see broad policy reforms for decentralization and devolution, in line with the directions given in the government's Eighth Five Year Plan," Saito said. "Certainly, the national urban policy is something ADB can look at supporting more."

Without a national urban policy, *pourashavas* may continue to rely on project-based opportunities for infrastructure improvements, capacity building, and social development support, explained Alexandra Vogl, UGIIP-3 additional financing mission leader and principal planning and policy specialist in the Strategy, Policy, and Business Process Division of ADB's Strategy, Policy, and Partnerships Department.

More livable cities. The UGIIP projects have built 82 new municipal facilities in *pourashavas*: schools, community centers, bus and truck terminals, street lighting, kitchen markets and slaughterhouses, shopping centers (or municipal markets), ghats (boat landing stations), and recreation centers. The facilities reduce inequalities and increase access to markets, goods, services and even quality education.

Pourashava-in-Focus: **Nilphamari**

Impact Story 17: Learning to Sew, Save, Build Dream Homes Raises "Revolutionary" Voices

In the impoverished area of Sawdagar Para, roads were in as bad of shape as people's livelihoods and income levels. Umme Habiba, 25, and her neighbors said they would often be trapped for months from waterlogging during the monsoon season. Environmental conditions make the hardship of poverty all the more difficult to overcome. In 2016, Habiba was forced to leave her college after the sudden death of her father—an enormous loss, she said.

A 1-month training program on sewing provided her the skills (as well as a sewing machine) to deal with her loss and its aftermath. "The training program changed my destiny," she said. "I completed my college with the daily income I earned by fulfilling people's sewing orders."

Since finishing college, Habiba has married and is now a mother. She earns Tk9,000–Tk12,000 monthly (about $104–$140), enough to also help her husband's family. "The biggest investment I made in the last 5 years," she said, "is saving for our house." Habiba's husband is slowly working on their newly constructed two-storey home built on ancestral land. Habiba said the family believes their new home, which was once a wild dream, is soon about to come true.

Nilphamari *Pourashava* is prioritizing women's development and self-reliance. Women have formed cooperative groups to address local issues. "In an area like ours, women hardly get involved in money matters," Habiba said. "Now we have a voice in society and in our own homes, which I believe is revolutionary."

Sewing revolution. Hundreds of women like Umme Habiba in Nilphamari *Pourashava* have taken up sewing after project-sponsored trainings skilled and inspired them to become profitable entrepreneurs.

Source: Authors.

APPENDIXES

A place to call home. Ruma Begum was granted a home in the Bangabandhu Pouro Abashon Prokolpo in Khagrachari *Pourashava* along with 33 other extremely vulnerable households with family members who are either disabled, aging veterans, or widowed.

APPENDIX 1
Key Documents

Asian Development Bank (ADB). 1990. *Report and Recommendation of the President to the Board of Directors: Proposed Loan and Technical Assistance Grant to the People's Republic of Bangladesh for the Secondary Towns Infrastructure Development Project*. Manila.

———. 1995. *Report and Recommendation of the President to the Board of Directors: Proposed Loan to the People's Republic of Bangladesh for the Secondary Towns Infrastructure Development Project II*. Manila.

———. 1997. *Technical Assistance to Bangladesh for the Third Urban Development Project*. Manila.

———. 2002. *Report and Recommendation of the President to the Board of Directors: Proposed Loan and Technical Assistance Grant to the People's Republic of Bangladesh for the Urban Governance and Infrastructure (Sector) Improvement Project*. Manila.

———. 2006. *Report and Recommendation of the President to the Board of Directors: Proposed Loan to the People's Republic of Bangladesh for the Secondary Towns Water Supply and Sanitation Sector Project*. Manila (Loan 2265-BAN [SF], for $41 million [Asian Development Fund], approved on 16 October).

———. 2012. *The Urban Governance and Infrastructure Improvement (Sector) Project in Bangladesh: Sharing Knowledge on Community-Driven Development*. Manila

———. 2012. *Project Completion Report: Urban Governance and Infrastructure Improvement (Sector) Project in Bangladesh*. Manila.

———. 2014. *Report and Recommendation of the President to the Board of Directors: Proposed Loan and Administration of Loan to the People's Republic of Bangladesh for the Third Urban Governance and Infrastructure Improvement (Sector) Project*. Manila.

———. 2015. Bangladesh: Transforming Communities, Changing Lives. In ADB. 2015. *Together We Deliver 2014: From Knowledge and Partnerships to Results*. Manila. pp. 12–20.

———. 2015. *Strengthening Municipal Governance through Performance-Based Budget Allocation in Bangladesh*. Manila.

———. 2017. *Project Completion Report: Second Urban Governance and Infrastructure Improvement (Sector) Project in Bangladesh*. Manila.

Government of Bangladesh, Planning Commission. 2020. *Eighth Five Year Plan, July 2020–June 2025*. Dhaka.

Government of Bangladesh, Local Government Engineering Department and Department of Public Health Engineering. 2022. *Our Talks: Success Stories of Self-Reliant Women in UGIIP-III Under LGED*. Dhaka.

APPENDIX 2
Key Interviews

Asian Development Bank

Rafiqul Islam, former senior project officer (urban infrastructure), in the Asian Development Bank (ADB) Bangladesh Resident Mission

Hun Kim, Urban Governance and Infrastructure Improvement (Sector) Project (UGIIP) mission leader and former director general of ADB's South Asia Department

Masayuki Tachiiri, UGIIP-2 mission leader and director of the Strategy, Policy, and Business Process Division of ADB's Strategy, Policy, and Partnerships Department

Norio Saito, UGIIP-3 mission leader and director of ADB's South Asia Urban and Water Division

Alexandra Vogl, UGIIP-3 mission leader for additional financing and principal planning and policy specialist in the Strategy, Policy, and Business Process Division of ADB's Strategy, Policy, and Partnerships Department

Local Government Engineering Department

AKM Rezaul Islam, project director, UGIIP-3

Md. Shafiqul Islam Akand, former UGIIP project director (2011–2015) and former additional chief engineer for the Local Government Engineering Department

Azahar Ali, team leader, Governance Improvement and Capacity Development Consultancy

Nilufar Yesmin, Jr., gender development and poverty reduction specialist

Water at home. After a lifetime of carrying water over long distances, 75-year-old Nur Jahan Begum gained a metered household water connection, due to Chapainawabganj *pourashava's* successful reforms earning infrastructure funding from ADB. Read more about her *pourashava* on page 40.

APPENDIX 3
The Evolution of Urban Governance Reform Indicators

Each Urban Governance and Infrastructure Improvement (Sector) Project (UGIIP) modified the Urban Governance Improvement Action Program, introducing new or varied indicators. UGIIP-3 introduced performance indicators that reflected the stage of governance development of project *pourashavas*: entry-level, intermediate, and advanced. UGIIP-3 also introduced a weighting system for the indicators. Appendixes 5 and 6 present UGIIP-3's more intricate indicators and weighting system.

Table A3: Comparison of the Urban Governance Improvement Action Plan between UGIIP-1, UGIIP-2, and UGIIP-3

Reform Areas	UGIIP-1 Indicators	UGIIP-2 Indicators	UGIIP-3 Indicators
1. Residents' awareness and participation	1.1 Build information system for the general public (e.g., publicity board in each ward; notices in local newspapers; semiannual leaflets) on the UGIAP and its activities	1.1 Resident charter approved by TLCC; displayed at *pourashava* office	1.1 Formation and operation of TLCCs[a]
	1.2 Form TLCC (within 6 months of signing subproject agreement); committee will consist of *pourashava* chairperson (as the chair), ward commissioners, representatives of civil society organizations, professional associations, and resident groups (including women); total number of committee members not to exceed 60	1.2 Resident report cards prepared, approved, and implemented by TLCC	1.2 Formation and working of WLCCs
	1.3 Regular meetings of standing committees held with at least one-third of the members attending	1.3 Grievance redress cell established, functional with clear terms of reference	1.3 Preparation and implementation of resident charter
	1.4 Establish planning unit supported by full-time town planning expert to be engaged under the project	1.4 TLCC, WLCC meetings held regularly	1.4 Formation and operation of information and grievance redress cell
	1.5 Implementation of infrastructure inventory assessment and mapping	1.5 Budget proposal compared with budget and actual outlays of previous year; displayed at the *pourashava* office; discussed at TLCC	
	1.6 Discussion, endorsement of municipal infrastructure development plan (complete with updated land use plan) to be prepared by planning unit with assistance from urban management support unit	1.6 Mass communication cell established, and campaign plan developed and implemented as planned	

continued on next page.

Table A3 continued.

Reform Areas	UGIIP-1 Indicators	UGIIP-2 Indicators	UGIIP-3 Indicators
	1.7 Finalization of implementation plan for municipal infrastructure development plan	1.7 Finalization of implementation plan for municipal infrastructure development plan	
Changes in indicators between UGIIP-1 and UGIIP-3:			
2. Women's participation	2.1 Municipal council in each *pourashava* delineates responsibilities of FWCs	2.1 GAP prepared, included in *pourashava* development plan	2.1 Form and activate standing committee on women and children (according to prescribed guidelines) to prepare and steer a customized GAP
	2.2 Formation of gender and environment subcommittee headed by an FWC	2.2 Budget to implement GAP identified, approved.	2.2 Form and activate standing committee on poverty reduction and slum improvement (based on guidelines) to prepare and steer a customized poverty reduction action plan
	2.3 Participation of FWCs in *pourashava* committees		2.3 Form slum improvement committee to implement slum improvement activities
	2.4 Organization of semiannual rallies on UGIIP; social and gender development at the ward level involving female residents		
	2.5 Activities undertaken in accordance with the GAP (based on targets)		
	2.6 Introduction of monitoring and report system on the GAP; (PMO will develop the system)		
Changes in indicators between UGIIP-1 and UGIIP-3:			
3. Integration of the urban poor	3.1 Preparation of *pourashava*-level poverty reduction action program	3.1 Slum improvement committees established in targeted slums	These governance indicators were integrated with indicators of women's participation; the combined indicator was titled "Equity and Inclusiveness of Women and Urban Poor"
	3.2 Formation of slum improvement committees in targeted slums	3.2 Poverty reduction action plan prepared and endorsed by TLCC; fully implemented; quarterly reports prepared	
	3.3 Organization of semiannual rallies and consultations on environment, water supply, sanitation, and solid waste management as envisaged in the GAP	3.3 Budget to implement poverty reduction action plan identified and approved	

continued on next page.

Table A3 continued.

Reform Areas	UGIIP-1 Indicators	UGIIP-2 Indicators	UGIIP-3 Indicators
	3.4 Engagement of nongovernment organizations for microcredit for poor and low-income women entrepreneurs and preparation of a fund utilization plan by the implementing nongovernment organization		
	3.5 Lending activities conducted by the implementing nongovernment organizations		
	3.6 Delivery of health and education programs in targeted slums as envisioned in the poverty reduction action plan		
	3.7 Leadership and skills training offered to poor women as envisaged in the poverty reduction action plan		
	3.8 Establish a poverty reduction action plan-based monitoring and reporting system		
Evolution of indicators between UGIIP-1 and UGIIP-3:			
4. Financial accountability and sustainability	4.1 Increase in the holding tax collection efficiency each year	4.1 Computerized accounting system introduced; computer-generated accounting reports produced	4.1 Preparation of annual *pourashava* budget with standing committee on establishment and finance
	4.2 Other own source collection	4.2 Computerized tax record system introduced; computer-generated bills produced	4.2 Carrying out audit of accounts with standing committee on accounts and audit
	4.3 Compliance with property tax appraisal (regular reassessment every 5 years)	4.3 Financial statements prepared; audit by standing committee on accounts and audit within 3 months of fiscal year closure	4.3 Establishing computerized accounting system; generating computerized accounting reports
	4.4 Regular payment of electricity bills	4.4 Interim tax assessment carried out annually; collection increased	4.4. Payment of electric and telephone bills
	4.5 Regular payment of telephone bills	4.5 Nontax own revenue source increased at least by inflation rate	4.5 Carrying out inventory of fixed assets, opening fixed asset register, designing fixed asset database, creating fixed asset depreciation fund account
	4.6 Budget reservations for O&M	4.6 All debts due to Government of Bangladesh and/or other entities fully repaid according to schedule; ratio of debt servicing to annual revenue receipts remains less than 25%	4.6 Repayment of all Government of Bangladesh loans

continued on next page.

Table A3 continued.

Reform Areas	UGIIP-1 Indicators	UGIIP-2 Indicators	UGIIP-3 Indicators
	4.7 Computerization, improved management of tax records with assistance of urban management support unit	4.7 All outstanding bills older than 3 months, including electricity and telephone, paid in full	
	4.8 Computerization, improved reporting of accounting records with assistance of urban management support unit		
Evolution of indicators between UGIIP-1 and UGIIP-3:			
5. Administrative transparency	5.1 Development of adequate staff structure (according to size and needs) with detailed job descriptions to enable *pourashava* to effectively undertake its current and future obligations	5.1 Development of adequate staff structure (according to size and needs) with detailed job descriptions to enable *pourashava* to effectively undertake current and future obligations	5.1 Formation and operation of standing committees (see Article 55 of the Local Government [*Pourashava*] Act of 2009)
	5.2 Establishment of ward-level coordinating committees headed by respective ward commissioners	5.2 Elected representatives, *pourashava* officials, and concerned residents actively participate in training programs	5.2 Ensure participation and assistance in conducting all training programs
	5.3 Activation of *pourashava* subcommittees	5.3 Progress report on UGIAP implementation and other activities submitted on time to the PMO	5.3 Use improved information technology for good governance (see Article 54 of the Local Government [*Pourashava*] Act of 2009)
	5.4 Participation in training programs of elected officials at urban management support unit	5.4 Standing committees established and/or activated	
	5.5 Evaluation and monitoring by LGED offices of the progress and quality of physical works	5.5 Ensure evaluation and monitoring by regional LGED offices on progress and quality of physical works	
	5.6 Delivery of progress reports to the PMO	5.6 Initiate e-governance activities	
Evolution of indicator between UGIIP-1 and UGIIP-3:			
6. Urban planning		6.1 Base map verified; updated land use plan prepared	6.1 Preparation and implementation of *pourashava* development plan
		6.2 Annual O&M plan, including budget requirement, prepared and approved as part of *pourashava* development plan	6.2 Control of development activities
		6.3 A full-time *pourashava* urban planner recruited (class A *pourashavas* only)	6.3 Prepare annual O&M plan, including budget provision

continued on next page.

Table A3 continued.

Reform Areas	UGIIP-1 Indicators	UGIIP-2 Indicators	UGIIP-3 Indicators
Evolution of indicator between UGIIP-1 and UGIIP-3:			
7. Enhancement of local resource mobilization			7.1 Revenue mobilization through holding tax
			7.2 Revenue mobilization through collection of indirect taxes and fees from sources other than holding tax
			7.3 Computerize tax record system; generate computerized tax bill
			7.4 Fixing and collection of water tariff
8. Keeping essential *pourashava* services functional			8.1 Solid waste collection, disposal, and management
			8.2 Drain cleaning and maintenance
			8.3 Arrangement for making street lighting functional
			8.4 Carrying out infrastructure O&M; establishment and operation of mobile maintenance team
			8.5 Managing sanitation

FWC = female ward commissioner, GAP = gender action plan, LGED = Local Government Engineering Department, O&M = operation and maintenance, PMO = project management office, TLCC = town-level coordinating committee, UGIAP = Urban Governance Improvement Action Program, UGIIP = Urban Governance and Infrastructure Improvement (Sector) Project, WLCC = ward-level coordinating committee.

Note:[a] By UGIIP-3, the Local Government (*Pourashava*) Act of 2009, Article 115, mandated that *pourashavas* establish TLCCs. Though *pourashavas* may have established TLCCs, UGIIP has helped *pourashavas* improve the operation and effectiveness of TLCCs.

Source: UGIIP Project Management Office.

More equitable markets. Khagrachari *Pourashava* utilized its UGIIP infrastructure funding to relocate its weekly market to provide easier access for indigenous communities that travel from distant villages to sell. Read more about this *pourashava* in a story on page 56.

APPENDIX 4
Urban Governance Reform Indicators for Window A *Pourashavas*

Each Urban Governance and Infrastructure Improvement (Sector) Project (UGIIP) modified the Urban Governance Improvement Action Program (UGIAP), introducing new or varied indicators. UGIIP-3 introduced performance indicators. Table A4 is a detailed presentation of UGIIP-3 UGIAP indicators, which were the most developed and elaborate of the three UGIIPs. As *pourashavas* progressed through each phase (entry, intermediate, and advanced), they received a proportion of their infrastructure allotment. The following guidelines apply when reading Table A4:

- For the entry criteria, all criteria must be fulfilled for phase 1 investment budget allocation.
- *Pourashavas* that fulfill core activities and score 80% or higher are considered "fully satisfactory" and will be entitled to the full funding for the next phase of the investment budget.
- *Pourashavas* that fulfill all core activities and score from 60% to 80% for noncore activities are considered "generally satisfactory" and will be entitled to partial funding for the next phase of the investment budget.
- *Pourashavas* that fulfill all core activities, but score less than 60% for noncore activities are considered "not satisfactory" and will not be entitled to the investment budget allocation.

Table A4: Urban Governance Improvement Action Plan Criteria for Window A *Pourashavas* under UGIIP-3

Area of Activity	Performance Indicator/Criteria			Activities and Weight of Non-Core Indicator Activities
	Entry	Intermediate	Advanced	
1. Resident Awareness and Participation				
(i) Formation and operation of TLCCs (Reference: Article 115 of the Local Government [*Pourashava*] Act of 2009)	• TLCC formed as per procedure • At least two meetings held • Meeting agenda and minutes prepared and disclosed	• Meetings held at regular intervals • Participation of all members, including women and the poor, in discussions ensured • Meeting working paper and minutes prepared and disclosed, and decisions followed up on	• Meetings held at regular intervals • Participation of all members, including women and the poor, in discussions ensured • Meeting working paper and minutes prepared and disclosed on *pourashava* website, and decisions followed up on	Core

continued on next page.

Table A4 continued.

Area of Activity	Performance Indicator/Criteria			Activities and Weight of Non-Core Indicator Activities
	Entry	Intermediate	Advanced	
(ii) Formation and operation of ward-level coordination committee (WLCC) (Reference: Article 14 of the Local Government [*Pourashava*] Act of 2009)	• WLCC formed as per procedure • At least one meeting held in each ward	• *Meetings held at regular intervals*[a] • *Participation of all members, including women and the poor, in discussions ensured* • *Meetings held and records kept and communicated to the pourashava*	• Meetings held at regular intervals • Participation of all members in discussions ensured • Meetings held and records kept and communicated to the pourashava	1
(iii) Preparation and implementation of resident charter (Reference: Article 53 of the Local Government [*Pourashava*] Act of 2009)		• *Resident charter prepared and endorsed by TLCC and pourashava council* • *Resident charter displayed in pourashava office and other important places, and prescribed services delivered*	• Charter continues to be displayed • Establish reception and service center at pourashava office	1
(iv) Formation and operation of information and grievance redress cell		• *Complaint box installed in pourashava office* • *Grievance redress cell formed as per procedures* • *Meetings held as and when required* • *Grievance redress cell activities disclosed to TLCC*	• Complaint box remains available • Meetings held as and when required • Meeting decisions communicated to complainants, and pourashava council informed • Grievance redress cell activities disclosed to TLCC and on pourashava website	2
2. Urban Planning				
(i) Preparation and Implementation of PDP	• PDP prepared through a participatory process • PDP endorsed by TLCC and approved by *pourashava* council	• *Development activities taken up conforming to the PDP*	• Development activities taken up conforming to the PDP	1
(ii) Control of development activities		• *Urban planning unit functional* • *Enforce at least 60% control of building construction, reconstruction activities, land development* • *Effective prevention of encroachment on public land (e.g., river, canal, khas land [government-owned fallow land]) practiced*	• Urban planning unit functional • Enforce at least 60% control of building construction, reconstruction activities, land development • Effective prevention of encroachment on public land (e.g., river, canal, khas land [government-owned fallow land]) practiced	3

continued on next page.

Table A4 continued.

Area of Activity	Performance Indicator/Criteria			Activities and Weight of Non-Core Indicator Activities
	Entry	Intermediate	Advanced	
(iii) Prepare annual O&M plan including budget provision		• O&M plan prepared, approved, implemented, and posted on pourashava website • Increased budget allocation by at least 5% each year	• O&M plan prepared, approved, implemented, and posted on pourashava website • Increased budget allocation by at least 5% each year	2
3. Equity and Inclusiveness of Women and Urban Poor				
(i) Form and activate standing committee on women and children (according to prescribed guidelines) to prepare and steer the customized GAP (Reference: Article 55 of the Local Government [*Pourashava*] Act of 2009)	Standing committee formed as per prescribed guidelines At least two meetings held Core activities under the GAP identified	• Meetings held at regular intervals, with agenda and minutes prepared and disclosed • GAP with activities and responsibilities endorsed by TLCC • GAP implementation takes place with allocated funds from revenue budget in accordance with the GAP	• Meetings held at regular intervals, with agenda and minutes prepared and disclosed • GAP implementation takes place with allocated funds from revenue budget (5% higher than the previous year) • GAP implementation report prepared and disclosed	2
(ii) Form and activate standing committee on poverty reduction and slum improvement (according to prescribed guidelines) to prepare and steer the customized PRAP (Reference: Article 55 of the Local Government [*Pourashava*] Act of 2009)	Standing committee formed as per prescribed guidelines At least two meetings held Core activities under PRAP identified	• Meetings held at regular intervals, with agenda and minutes prepared and disclosed • PRAP with activities and responsibilities endorsed by TLCC • PRAP implementation taking place with allocated funds from revenue budget in accordance with the plan	• Meetings held at regular intervals, with agenda and minutes prepared and disclosed • PRAP implementation takes place with allocated fund from revenue budget (5% higher than the previous year) • PRAP implementation report prepared and disclosed	2
(iii) Form SICs to implement slum improvement activities	Slum selection done according to priority	• SICs formed in selected slums • Regular meetings of SICs held • Slum improvement activities implemented by SICs with effective participation of all members	• Regular meetings held • Slum improvement activities implemented by SICs with effective participation of all members	1

continued on next page.

Table A4 continued.

Area of Activity	Performance Indicator/Criteria			Activities and Weight of Non-Core Indicator Activities
	Entry	Intermediate	Advanced	
4. Enhancement of Local Resource Mobilization				
(i) Revenue mobilization through holding tax	Action plan for enhanced holding tax endorsed by TLCC[b]	• Regular assessment done at 5-year intervals if due, and interim assessment done every year as per rule or procedure • Increased holding tax collected, including arrears (at least 70% of demand)	• Regular assessment done at 5-year intervals if due, and interim assessment done every year as per rule or procedures • Increased holding tax collected, including arrears (at least 80% of demand) • Actions initiated against major defaulters	Core
(ii) Revenue mobilization through collection of indirect taxes and fees from other sources (other than holding tax)	Action plan for enhanced tax revenue and endorsed by TLCC	• The amount of indirect taxes, fees, rentals, and lease money charged and collected, including arrears, increased annually by at least 5%	• The amount of indirect taxes, fees, rentals, and lease money charged and collected, including arrears, increased annually by at least the official inflation rate	Core
(iii) Computerize tax record system and generate computerized tax bill		• Computerized tax record software installed, and database prepared • Computerized tax bill generated and served to customers	• Computerized database updated • Computerized tax bill generated and served to customers	1
(iv) Fixing and collection of water tariff	Tariff enhancement plan prepared Commitment by *pourashava* to install water meter and subsequently collect tariff on volumetric consumption obtained (where applicable)	• Tariff enhancement plan implemented • Inventory of assets prepared and published • Water bills collection through banks initiated	• Tariff collection efficiency of at least 80% achieved • Inventory of assets updated and published • Action initiated for introducing volumetric water tariff • Water tariff collected through computerized system and banks	3

continued on next page.

Table A4 continued.

Area of Activity	Performance Indicator/Criteria			Activities and Weight of Non-Core Indicator Activities
	Entry	Intermediate	Advanced	
5. Financial Management, Accountability, and Sustainability				
(i) Preparation of annual *pourashava* budget with involvement of standing committee on establishment and finance (Reference: Article 55 of the Local Government [*Pourashava*] Act of 2009)	Annual budget approved and disclosed	• **Estimated budget modified based on comments and suggestions from residents and TLCC** • **Annual budget approved by *pourashava* council**	• **Estimated budget modified based on comments and suggestions from residents and TLCC** • **Annual budget approved by *pourashava* council and posted on *pourashava* website**	1
(ii) Carrying out of audit of accounts with involvement of standing committee on accounts and audit (Reference: Article 55 of the Local Government [*Pourashava*] Act of 2009)		• **Annual statement of income and expenditure prepared** • **Audit conducted by standing committee on accounts and audit once a year and report prepared** • **Audit report of the standing committee presented to TLCC and *pourashava* council and sent to the PMO within 3 months**	• **Annual statement of income and expenditure prepared** • **Audit conducted by standing committee on accounts and audit once a year and report prepared** • **Audit report of the standing committee presented to TLCC and *pourashava* council, and posted on *pourashava* website and sent to the PMO within 3 months**	Core
(iii) Establishment of computerized accounting system and generation of computerized accounting reports		• *Computerized accounting system installed*	• *Computerized accounting reports generated*	1
(iv) Payment of electric and telephone bills	• **Plan prepared for clearing arrears, if any, of electric and telephone bills**	• **Electric and telephone bills (current and in arrears) paid (80% of total bills and certificates obtained from concerned authority)**	• **Electric and telephone bills (current ad in arrears) paid (90% of total bills and certificates obtained from concerned authority)**	Core
(v) Carrying out of inventory of fixed assets, opening of fixed asset register, designing of fixed asset database, and creation of fixed asset depreciation fund account	• **Inventory of fixed assets done** • **Fixed asset register opened and used**	• *Inventory of fixed assets updated* • *Rental and lease values of property updated and increased* • *Fixed asset database installed and used*	• *Inventory of fixed assets updated* • *Rental and lease values of property regularly updated and increased* • *Use of fixed asset database continued* • *Fixed asset depreciation fund account created*	2

continued on next page.

Table A4 continued.

Area of Activity	Performance Indicator/Criteria			Activities and Weight of Non-Core Indicator Activities
	Entry	Intermediate	Advanced	
(vi) Repayment of all Government of Bangladesh loans	Plan prepared for clearing the overdue amount, if any, of outstanding loans	• At least 80% of all Government of Bangladesh and BMDF loan repaid as scheduled, and unpaid amount rescheduled	• At least 90% of all Government of Bangladesh and BMDF loan repaid as scheduled, and unpaid amount rescheduled	2
6. Administrative Transparency				
(i) Formation and operation of standing committees (Reference: Article 55 of the Local Government [*Pourashava*] Act of 2009)	All standing committees formed as per procedure At least one meeting held for each standing committee	• Standing committee meetings held at prescribed intervals • Meeting agenda and minutes prepared and disclosed to TLCC	• Standing committee meetings held at prescribed intervals • Meeting agenda and minutes prepared and disclosed to TLCC	2
(ii) Ensure participation and assistance in conducting all training programs		• Participation in all training programs ensured • Training program from own pourashava budget planned and implemented	• Participation in all training programs ensured • Training program from own pourashava budget planned and implemented	1
(iii) Using improved information technology for good governance (Reference: Article 54 of the Local Government [*Pourashava*] Act of 2009)		• Pourashava website activated and maintained • All relevant information uploaded and regularly updated	• Pourashava website activated and maintained • All relevant information uploaded and regularly updated	2
7. Keeping Essential *Pourashava* Services Functional				
(i) Collection, disposal, and management of solid waste	**Action plan prepared**	• Action plan implemented with budget allocation • Regular collection done in core areas • TLCC's satisfaction level assessed	• Action plan implemented with budget allocation • Regular collection done in core area and solid waste disposed of at a safe site (or at least show progress is on track) • TLCC's satisfaction level assessed	Core
(ii) Cleaning and maintenance of drains	**Action plan prepared**	• Action plan implemented with budget allocation • Regular cleaning of primary drains done • TLCC's satisfaction level assessed	• Action plan implemented with budget allocation • Regular cleaning of primary and secondary drains done • TLCC's satisfaction level assessed	Core

continued on next page.

Table A4 continued.

Area of Activity	Performance Indicator/Criteria			Activities and Weight of Non-Core Indicator Activities
	Entry	Intermediate	Advanced	
(iii) Arrangement for making street lighting functional	**Action plan prepared**	• Action plan implemented with budget allocation • Street lighting functional along 80% of streets • TLCC's satisfaction level assessed	• Action plan implemented with budget allocation • Street lighting functional along 90% of streets • TLCC's satisfaction level assessed	Core
(iv) Carrying out O&M of infrastructure, and establishment and operation of mobile maintenance team	**Action plan prepared**	• Action plan implemented with budget allocation priority • O&M activities implemented • Mobile maintenance team functional • TLCC's satisfaction level assessed	• Action plan implemented with budget allocation • O&M activities fully operational • Mobile maintenance team functional • TLCC's satisfaction level assessed	Core
(v) Managing sanitation	**Action plan prepared**	• **Action plan implemented with budget allocation** • **Public toilets made functional and cleaned** • **TLCC's satisfaction level assessed**	• **Annual program and budget prepared** • **Public toilets made functional and cleaned** • **Fecal sludge management initiated** • **TLCC's satisfaction level assessed**	Core

BMDF = Bangladesh Municipal Development Fund, GAP = gender action plan, O&M = operation and maintenance, PDP = *pourashava* development plan, PMO = project management office, PRAP = poverty reduction action plan, SIC = slum improvement committee, TLCC = town-level coordinating committee, UGIIP = Urban Governance and Infrastructure Improvement (Sector) Project, WLCC = ward-level coordinating committee.

Notes:
a For non-core activities/criteria under the intermediate and advanced criteria shown in *italics*, the score will be given for each area of activity. The score of each activity is the weight given to each area of activity divided by the number of activities in each area (e.g., if there are three bullet points in an area of activity with a weight of 2, each activity has a score of 0.66 (2 divided by 3). It is either "pass" or "fail" for each activity.
b For the intermediate and advanced criteria, all core activities and criteria shown in **bold** must be fulfilled. If a *pourashava* fails to meet a single activity under the core areas of activity, it will not be entitled to the budget allocation for the next phase.

Source: Authors.

Computer training. Roksana Nazneen is a computer trainer at the Maulavibazar Computer Lab and is among the first batch of training graduates.

APPENDIX 5
Urban Governance Reform Indicators for Window B *Pourashavas*

Pourashavas that participated in the original Urban Governance and Infrastructure Improvement (Sector) Project (UGIIP) and UGIIP-2 had the opportunity to access up to $2 million more in development funds from UGIIP-3 if they met intermediate urban governance indicators. The intermediate indicators built from previous reforms achievements (Table A5).

Table A5: Urban Governance Indicator Action Plan for Window B *Pourashavas*

Activity	Task	Performance Indicators/Criteria	Assessment Method	Maximum Score
1. Resident Awareness and Participation				
1.1 Formation and working of TLCC and WLCC	1. Form TLCC 2. Hold TLCC meetings and prepare minutes 3. Form ward committees 4. Hold ward committee meetings	TLCCs and ward committees formed in accordance with the requirements	4 or 0	10
		• Meetings held at regular intervals	3 or 0	
		• Meeting minutes of TLCC prepared and disclosed	3 or 0	
2. Equity and Inclusiveness of Women and Urban Poor				
2.1 Planning and implementation of activities for women and urban poor	1. Identify activities 2. Allocate fund 3. Implement activities	Activities identified	2 or 0	10
		Fund allocated from *pourashava* budget	3 or 0	
		Percentage of fund allocated actually utilized	[Actual[a] percentage of a *pourashava*/ percentage of the highest-performing *pourashava*] x 5	
3. Enhancement of Local Resource Mobilization				
3.1 Revenue mobilization through holding tax	1. Conduct regular and interim assessment 2. Collect regular and interim holding tax, including arrears	• Regular assessment done at 5-year intervals if due	4 or 0	8
		• Interim assessment done on a continuous basis	4 or 0	
		• At least 70% of holding tax collected, including arrears	[Actual[b] percentage of a *pourashava* – 50%)/100%] x 2 x 12	12
3.2 Revenue mobilization through collection of indirect taxes and fees (other than holding tax)	1. Charge and collect indirect taxes and fees	• Increased indirect taxes and fees charged and collected, including arrears (at least 80% collection against demand, and collection increased by a minimum 7% each year)[c]	[Actual[d] % of a *pourashava* – 50%)/100] x 2 x 10	10

continued on next page.

Table A5 continued.

Activity	Task	Performance Indicators/Criteria	Assessment Method	Maximum Score
4. Financial Management, Accountability, and Sustainability				
4.1 Preparation of annual *pourashava* budget	1. Prepare annual budget 2. Obtain comments and/or suggestions from residents and endorsement of TLCC 3. Approved budget reaches LGD	Estimated budget disclosed to open public meetings and modified based on comments and/or suggestions from residents and TLCC	6 or 0	8
		Approved budget reaches LGD	2 or 0	

LGD = local government division, TLCC = town-level coordinating committee, WLCC = ward-level coordinating committee.

Notes:
[a] Maximum score to be granted to the highest-performing *pourashava*.
[b] Maximum score to be granted to the highest-performing *pourashava* = 12; achievement below 70% will result in a score of 0.
[c] Base year = 2015.
[d] Maximum score to be granted to the highest-performing *pourashava* = 10; achievement below 80% will result in a score of 0; *pourashavas* not meeting the minimum 7% annual increase will receive a score of 0. If no audit objections are issued, the *pourashava* will receive the full score. *Pourashavas* having no water supply component will be assessed based on a total point system of 94 (which will be subsequently multiplied by 100/94 to allow comparison with other *pourashavas* that will be scored on a total point system of 100).

Source: Urban Governance and Infrastructure Improvement (Sector) Project Management Office.

Headmaster. Masurda Begum Moni is the headmaster of Amir Uddin Pauro Primary School, a school built by Bhairab *Pourashava* with UGIIP funding. The school opened in 2010 and currently has 300 students enrolled from the low-income community it serves. Prior to UGIIP, the community did not have a nearby school.

APPENDIX 6
Benapole, a Border Town, Grows with UGIIP Support

Only 80 kilometers away from India's crucial commercial city of Kolkata, on the Bangladesh side of the border with India, the town of Benapole is what developers might call "booming." Urbanization began only in the 1990s in Benapole, leading the central government to reclassify Benapole in 2011 from a class C *pourashava* to a class A. The upgrade reflects its strategic economic importance and potential.

As Bangladesh's largest land port city, Benapole has been called a "gateway to Bangladesh." The Benapole land port is the most critical land port of Bangladesh and is operated by the Bangladesh Land Port Authority. About 90% of imported Indian goods enter Bangladesh through the Benapole Land Port, operated by the Bangladesh Land Port Authority. Geographically, Benapole is a major strategic point for border trading between Bangladesh and India, owing to its proximity to Kolkata. According to the Bangladesh Land Port Authority, about 90% of all imported items from India come through Benapole.

The goods that move across the border into Benapole are of national importance, but the jobs and people that move with those goods are Benapole's responsibility. In 2011, Ashraful Alam Liton was elected the *pourashava's* first official mayor. He was 33 years old in 2011 and was a political newcomer to elected office.

A gateway city. Benapole is the largest land port city in Bangladesh and has been called a "gateway to Bangladesh." The *pourashava* spent UGIIP funds on a community center that it used during the COVID-19 pandemic for laborers in transit to quarantine.

Full hands. Women take on childcare and livelihood training provided by Benapole *Pourashava*, which began the trainings as part of a UGIIP-initiative to reduce poverty and empower women economically and socially.

He had been a social worker and political operative. Getting elected was one achievement, but governing was another challenge. "I had no idea what to do once I was elected. I had no real way of connecting with the people on a citywide scale. There was a real disconnect," he said.

Being a newly declared official city, Benapole did not yet have an official office, skilled staff, or the financial resources for public services. "We were starting from scratch," Liton said. "We had absolutely nothing."

Benapole was selected to participate in the second Urban Governance and Infrastructure Improvement (Sector) Project (UGIIP-2), financed by the Asian Development Bank and executed by the Local Government Engineering Department and Department of Public Health Engineering. *Pourashavas* participating in UGIIP are eligible to earn investment finance for local infrastructure projects by achieving milestones in improved governance, such as ensuring resident participation in government decision-making, developing and implementing local poverty reduction and gender development programs, generating local revenues through effective taxation and fees, and conducting effective operation and maintenance of essential public infrastructure and services. As *pourashavas* progress through three phases of reforms, an independent government committee evaluates their performance. If their performance is satisfactory, *pourashavas* receive their allocated investment funds. Not meeting the fundamental governance indicators disqualifies them from participating in the project. The governance criteria become more demanding the further a *pourashava* progresses with the project as a strategy for ensuring continuous improvements.

Benapole was rated "fully satisfactory" for its performance in the first two phases of UGIIP-2, qualifying it for its full allotment of project funding, Tk100 million. It received Tk175 million ($2.2 million) because of its upgrade to a class A *pourashava* and performed even better in the third phase of the UGIIP-2. One of the significant works Benapole invested project funds in was 2,844 meters of drainage, with footpaths, in two locations of the city.

By implementing an effective tax assessment, billing, and collection system, Benapole increased its tax collection rate to 89.35% in 2021. Before UGIIP-2, Benapole did not collect revenues other than from local property taxes. Through UGIIP-2, Benapole developed nontax income streams from land transfers, bus terminal fees, trade licenses, water tariffs, and community center rentals. Nontax revenue collection increased from 61.5% in 2009 to 101.02% in 2021. Benapole collects an average of Tk15.23 million (about $179,000) annually from tax and TK26.39 million (about $310,948.9) in nontax revenues.

The increase in local revenues has supported the establishment and expansion of basic infrastructure and public services. For example, Benapole has introduced piped water supply to about 5,000 connections with a new project already approved for expanded coverage. All households have connections to the electrical grid. Benapole *Pourashava* has begun developing its fixed assets with a compost plant, a bus terminal, a truck terminal, and a community center. The *pourashava* has also built a shelter for street sweepers and other low-income laborers.

Benapole is one of the very few *pourashavas* to have developed and secured central government approval for its master plan. Liton says the *pourashava* is complying with it, though some of the envisioned development is physically challenging because of existing unplanned development and systems of narrow streets. Addressing some of the development issues will require collaboration with communities. The town-level coordinating committee (TLCC) is instrumental in brainstorming solutions, Liton said. The TLCC is a mixed government and resident group established in compliance with UGIIP governance requirements, but is now required by the new Local Government (*Pourashava*) Act of 2009, another accomplishment of UGIIP. "UGIIP has helped in avoiding confrontations with the public because we work so closely with the public, the people," Liton said. "The UGIIP model makes politicians work for the public."

> *"Formal education is not valued here because job opportunities are high. Education is neglected."*
>
> **Ashraful Alam Liton**
> mayor of Benapole *Pourashava*

Because of its strong performance in UGIIP-2, Benapole also participated in UGIIP-3. (UGIIP-2 was implemented from 2008 to 2016. UGIIP-3 was implemented in 2014 and is ongoing.) Still the mayor of Benapole, Liton has two major priorities that will need public support and funding. One of these is a modern health care facility, not just for local residents, but also for the transient labor population and those working on the 500 trucks that come from India daily and pass through Benapole for loading and unloading goods. "They also have health issues, and not treating them will affect the local health standards," Liton said. The lack of a health care facility and the presence of migrant labor became a serious concern for Benapole during the coronavirus disease (COVID-19) pandemic. Liton said the *pourashava* converted its

Raising awareness, livelihoods. Benapole *Pourashava* holds a community meeting to share information about upcoming livelihood training programs.

new community center into a quarantine center for stranded laborers who crossed the Bangladesh–India border regularly. Liton hopes Benapole's development will attract investments in recreational and hospitality services, which are lacking. Benapole needs hostels and hotels for the different types of laborers and travelers that the border area attracts. Liton also wants to develop a quality college in Benapole. "Formal education is not valued here because job opportunities are plentiful. Education is neglected," he said. He worries that the local youth are passing up opportunities for higher-paying jobs that require higher education, but offer a better quality of life than the immediately available lower-wage jobs.

Liton credits UGIIP for his enthusiasm, the *pourashava's* financial capacity, and the public's engagement with the town's plans. The project made possible much of what the *pourashava* has accomplished since he became mayor in 2011. He would not call UGIIP a project, though. He calls it a "philosophy" that makes leaders accountable and responsible to ensure better service delivery to the public.

APPENDIX 7
Estimated Investment Required Versus Project Allocation

Table A7 lists the initial 36 *pourashavas* included in the Third Urban Governance and Infrastructure Improvement (Sector) Project from 2014 to 2017 under window A. All *pourashavas* are class A in the government's classification of economic size. The results are based on March 2022 data. The monetary amounts are taka, in units of lakh (or 100,000); Tk100,000 is equivalent to $0.0113 million.

Table A7: *Pourashava* Development Plan Budgets, Total Versus Allocated

No.	*Pourashava* Name	Estimated Cost of Initial *Pourashava* Development Plan (2014) (Tk100,000)[a]	Expected Actual Allocation for Second Revised *Pourashava* Development Plan Proposal (Tk100,000)[b]	Difference between Estimated Initial Cost of *Pourashava* Development Plan and Expected Actual Allocation (%)[c]	History of *Pourashava* Participation in UGIIP[d]
1	Bandarban	7,500.00	7,972.09	106	UGIIP-1, UGIIP-2, UGIIP-3
2	Benapole	19,771.73	6,950.37	35	UGIIP-2, UGIIP-3
3	Bera	25,197.64	6,941.40	28	UGIIP-3
4	Chapainawabgonj	20,557.70	7,626.16	37	UGIIP-1, UGIIP-3
5	Charghat	10,014.10	4,689.33	47	UGIIP-3
6	Chatak	4,199.80	5,549.39	132	UGIIP-3
7	Chuadanga	15,000.00	6,565.05	44	UGIIP-2, UGIIP-3
8	Cox's Bazar	32,618.26	14,792.40	45	UGIIP-2, UGIIP-3
9	Faridpur	34,472.01	15,369.17	45	UGIIP-2, UGIIP-3
10	Gopalgonj	100,867.20	16,479.82	16	UGIIP-2, UGIIP-3
11	Habigonj	7,500.00	5,137.10	68	UGIIP-1, UGIIP-3
12	Ishwardi	7,500.00	6,824.27	91	UGIIP-1, UGIIP-3
13	Jashore	22,223.71	10,029.88	45	UGIIP-3
14	Joypurhat	19,980.00	6,752.75	34	UGIIP-1, UGIIP-3
15	Khagrachari	7,500.00	9,701.61	129	UGIIP-1, UGIIP-2, UGIIP-3
16	Kishoreganj	4,500.00	8,692.95	193	UGIIP-1, UGIIP-3
17	Kotalipara	13,600.00	5,890.39	43	UGIIP-3
18	Kushtia	43,867.00	14,874.74	34	UGIIP-1, UGIIP-2, UGIIP-3
19	Laksam	7,432.00	10,946.47	147	UGIIP-1, UGIIP-3
20	Lakshmipur	7,500.00	8,012.23	107	UGIIP-1, UGIIP-3

Table A7 continued.

No.	Pourashava Name	Estimated Cost of Initial Pourashava Development Plan (2014) (Tk100,000)[a]	Expected Actual Allocation for Second Revised Pourashava Development Plan Proposal (Tk100,000)[b]	Difference between Estimated Initial Cost of Pourashava Development Plan and Expected Actual Allocation (%)[c]	History of Pourashava Participation in UGIIP[d]
21	Lalmonirhat	11,415.66	7,698.44	67	UGIIP-1, UGIIP-3
22	Magura	17,012.69	8,866.68	52	UGIIP-3
23	Meherpur	14,959.59	5,244.11	35	UGIIP-1, UGIIP-3
24	Moulvibazar	7,500.00	8,051.65	107	UGIIP-1, UGIIP-3
25	Muktagacha	11,938.75	4,470.33	37	UGIIP-3
26	Mymenshing[e]	44,268.86	11,970.83	27	UGIIP-2, UGIIP-3
27	Nabinagar	5,999.20	5,456.27	91	UGIIP-3
28	Naogaon	29,675.99	7,604.38	26	UGIIP-3
29	Netrokhona	4,499.50	7,752.56	172	UGIIP-1, UGIIP-3
30	Nilphamari	7,500.00	7,310.64	97	UGIIP-2, UGIIP-3
31	Panchagarh	7,500.00	5,001.54	67	UGIIP-1, UGIIP-3
32	Rajbari	11,078.49	7,228.75	65	UGIIP-1, UGIIP-3
33	Rangamati	7,491.00	7,660.39	102	UGIIP-1, UGIIP-3
34	Shahjadpur	13,268.00	3,601.47	27	UGIIP-1, UGIIP-3
35	Sherpur	12,320.69	9,977.04	81	UGIIP-1, UGIIP-3
36	Tungipara[f]	12,000.00	6,849.03	57	UGIIP-3
Total		**630,229.57**	**294,541.68**	**42**	

tk = taka, UGIIP = Urban Governance and Infrastructure Improvement (Sector) Project.

Notes:
[a] Pourashava development plan estimates were indicative for selected subprojects and not based on engineering estimates. For project support only. Pourashava development plan estimates considered six components only: transport, drainage, slum, sanitation, municipal facilities, and water supply.
[b] Revised estimate based on the latest subprojects under implementation. Actual costs would be close to the contract amount. Actual figures would be known after the works are completed.
[c] Percentage of project's support against the initial requirements at the beginning of the project.
[d] UGIIP's support. Only four pourashavas (Bandarban, Jhalokathi [window B], Khagrachari, and Kushtia) took part in all three UGIIPs, but none completed the full nine phases (three each for UGIIP-1, UGIIP-2, and UGIIP-3) during the implementation period.
[e] Mymensingh Pourashava was upgraded to a city corporation and excluded from the Urban Governance Improvement Action Program in late 2018.
[f] Tungipara Pourashava is 100% financed by the Government of Bangladesh.

Source: Government of Bangladesh, Local Government Engineering Department.

Small help, big returns. Kanchi Rani Das, 35, belongs to a minority community of Bhairab *Pourashava*. As part of the UGIIP-1 project, she and other women living in the slum received sewing training and micro-loans to establish businesses.

APPENDIX 8
Budget Reporting on Standing Committee on Women and Children

The tables in Appendix 8 demonstrate how the Third Urban Governance and Infrastructure Improvement (Sector) Project (UGIIP-3) has monitored and reported on the budget allocations and expenditures of the standing committee on women and children. The UGIIP strategy shows concern for the entire development process by requiring, year to year during implementation, that the standing committees act to develop action plans and that the *pourashavas* allocate a percentage of local revenues and show proof of disbursement. According to UGIIP-3 staff, by the end of 2021, the 35 project *pourashavas* had disbursed more than $4.17 million from own-budget sources to support the implementation of their gender action plans.

Table A8.1: Status of Formation and Budgetary Activities for the Standing Committee on Women and Children in 35 *Pourashavas*, January–December 2020

UGIIP-3 Region	No.	Name of *Pourashava*[a]	Formation Date of GAP	Meeting Held at Regular Intervals with Agenda	Meeting Minutes Prepared and Disclosed	Current Budget Allocation, 2020 (Tk)	Total Expenditure from Revenue Budget, 2020 (Tk)	Efficiency of Expenditure	Whether GAP Implemented, Disclosed
Bogura	1	Bera	11 May 14	Yes	Yes	1,707,958	1,253,072	73.37	Yes
	2	Chapainawabgonj	28 Jun 15	Yes	Yes	2,435,000	2,436,195	100.05	Yes
	3	Charghat	30 Apr 14	Yes	Yes	800,000	822,000	102.75	Yes
	4	Ishwardi	17 Jun 15	Yes	Yes	1,363,250	1,530,213	112.25	Yes
	5	Joypurhat	13 May 15	Yes	Yes	1,583,113	1,795,378	113.41	Yes
	6	Lalmonirhat	22 Oct 14	Yes	Yes	1,875,000	1,833,880	97.81	Yes
	7	Noagaon	23 Oct 14	Yes	Yes	1,618,000	1,141,180	70.53	Yes
	8	Nilphamari	22 Jun 14	Yes	Yes	1,595,990	3,024,690	189.52	Yes
	9	Panchagarh	5 May 15	Yes	Yes	2,050,000	2,943,910	143.61	Yes
	10	Shahjadpur	29 Aug 13	Yes	Yes	809,890	329,550	40.69	Yes
Cumilla	11	Bandarban	13 Sep 15	Yes	Yes	1,673,000	1,686,000	100.78	Yes
	12	Khagrachari	6 Sep 15	Yes	Yes	2,050,000	1,639,455	79.97	Yes
	13	Laksam	11 Jun 15	Yes	Yes	5,550,000	4,212,595	75.90	Yes
	14	Laxshmipur	27 Feb 14	Yes	Yes	2,575,000	2,725,518	105.85	Yes
	15	Nabinagar	24 Jun 15	Yes	Yes	718,359.5	741,530	103.23	Yes
	16	Rangamati	14 Jun 15	Yes	Yes	2,000,000	4,193,327	209.67	Yes
Magura	17	Benapole	11 Jun 15	Yes	Yes	802,500	550,095	68.55	Yes
	18	Chuadanga	25 May 15	Yes	Yes	1,514,973	1,281,574	84.59	Yes

continued on next page.

Table A8.1 continued.

UGIIP-3 Region	No.	Name of *Pourashava*[a]	Formation Date of GAP	Meeting Held at Regular Intervals with Agenda	Meeting Minutes Prepared and Disclosed	Current Budget Allocation, 2020 (Tk)	Total Expenditure from Revenue Budget, 2020 (Tk)	Efficiency of Expenditure	Whether GAP Implemented, Disclosed
	19	Jashore	30 Apr 15	Yes	Yes	3,600,000	2,697,465	74.93	Yes
	20	Kotalipara	30 Mar 14	Yes	Yes	552,300	362,275	65.59	Yes
	21	Magura	3 Dec 14	Yes	Yes	1,953,500	916,345	46.91	Yes
	22	Meherpur	24 Feb 14	Yes	Yes	958,940	808,500	84.31	Yes
	23	Rajbari	21 May 15	Yes	Yes	850,000	554,220	65.20	Yes
	24	Tungipara	30 Mar 14	Yes	Yes	790,172	188,900	23.91	Yes
Mymensingh	25	Chatak	28 Sep 14	Yes	Yes	815,000	263,350	32.31	Yes
	26	Habigonj	27 Feb 14	Yes	Yes	1,525,000	453,685	29.75	Yes
	27	Kishoreganj	16 Jul 15	Yes	Yes	2,395,000	898,312	37.51	Yes
	28	Moulvibazar	16 Apr 15	Yes	Yes	1,416,254	1,301,349	91.89	Yes
	29	Muktagacha	21 Oct 14	Yes	Yes	950,000	596,608	62.80	Yes
	30	Netrokona	23 Apr 14	Yes	Yes	1,855,000	1,081,540	58.30	Yes
	31	Sherpur	15 Dec 14	Yes	Yes	1,755,000	1,970,133	112.26	Yes
Additional Financing	32	Cox's Bazar	24 Dec 17	Yes	Yes	4,881,500	10,521,000	215.53	Yes
	33	Faridpur	26 Dec 17	Yes	Yes	4,750,000	4,735,131	99.69	Yes
	34	Gopalgonj	Not availble	Yes	Yes	1,617,480	399,000	24.67	Yes
	35	Kushtia	31 Dec 17	Yes	Yes	3,810,000	2,845,860	74.69	Yes
Total						**67,197,181**	**64,733,835**	**96.33**	

GAP = gender action plan, tk = taka, UGIIP = Urban Governance and Infrastructure Improvement (Sector) Project.
Note: [a] Mymensingh was not considered, as it was dropped from the project after it converted to a city corporation.
Source: UGIIP Project Management Office.

Table A8.2: Status of Formation and Activating of Standing Committee on Poverty Reduction and Slum Improvement in 35 *Pourashavas*, January–December 2020

UGIIP-3 Region	No.	Name of *Pourashava*[a]	Meeting Held at Regular Intervals with Agenda	Meeting Minutes Prepared and Disclosed	Budget Allocation, 2020 (Taka)	Total Expenditure, 2020 (Taka)	Efficiency of Expenditure (%)	Whether PRAP Implemented and Disclosed
Bogura	1	Bera	Yes	Yes	3,288,351	3,695,828	112.39	Yes
	2	Chapai	Yes	Yes	6,050,000	3,760,192	62.15	Yes
	3	Charghat	Yes	Yes	1,225,000	1,135,200	92.67	Yes
	4	Ishwardi	Yes	Yes	3,408,050	2,867,041	84.13	Yes
	5	Joypurhat	Yes	Yes	3,834,092	4,208,691	109.77	Yes
	6	Lalmonirhat	Yes	Yes	3,375,000	2,993,480	88.70	Yes
	7	Noagaon	Yes	Yes	4,171,000	3,059,089	73.34	Yes
	8	Nilphamari	Yes	Yes	2,982,250	4,255,030	142.68	Yes
	9	Panchagarh	Yes	Yes	3,100,000	4,691,230	151.33	Yes
	10	Shahjadpur	Yes	Yes	2,025,450	659,910	32.58	Yes
Cumilla	11	Bandarban	Yes	Yes	2,926,350	2,940,700	100.49	Yes
	12	Khagrachari	Yes	Yes	26,500,000	5,146,504	19.42	Yes
	13	Laksam	Yes	Yes	8,100,000	6,809,000	84.06	Yes
	14	Laxshmipur	Yes	Yes	5,738,500	3,881,419	67.64	Yes
	15	Nabinagar	Yes	Yes	1,770,899	1,557,602	87.96	Yes
	16	Rangamati	Yes	Yes	4,550,000	4,626,327	101.68	Yes
Magura	17	Benapole	Yes	Yes	1,850,000	2,502,679	135.28	Yes
	18	Chuadanga	Yes	Yes	3,951,000	2,452,420	62.07	Yes
	19	Jashore	Yes	Yes	8,625,000	7,997,134	92.72	Yes
	20	Kotalipara	Yes	Yes	1,380,750	649,500	47.04	Yes
	21	Magura	Yes	Yes	3,247,000	1,380,277	42.51	Yes
	22	Meherpur	Yes	Yes	2,397,350	1,726,000	72.00	Yes
	23	Rajbari	Yes	Yes	1,600,000	638,500	39.91	Yes
	24	Tungipara	Yes	Yes	1.212,788	233,240	19.23	Yes
Mymensingh	25	Chatak	Yes	Yes	1,925,000	1,025,300	53.26	Yes
	26	Habigonj	Yes	Yes	3,850,000	552,183	14.34	Yes
	27	Kishoreganj	Yes	Yes	5,962,500	2,166,975	36.34	Yes
	28	Moulvibazar	Yes	Yes	3,407,732	2,358,088	69.20	Yes
	29	Muktagacha	Yes	Yes	2,350,000	933,075	39.71	Yes
	30	Netrokona	Yes	Yes	4,625,000	4,460,027	96.43	Yes
	31	Sherpur	Yes	Yes	4,387,000	5,103,404	116.33	Yes

continued on next page.

Table A8.2 continued.

UGIIP-3 Region	No.	Name of Pourashava[a]	Meeting Held at Regular Intervals with Agenda	Meeting Minutes Prepared and Disclosed	Budget Allocation, 2020 (Taka)	Total Expenditure, 2020 (Taka)	Efficiency of Expenditure (%)	Whether PRAP Implemented and Disclosed
Additional Financing	32	Cox's Bazar	Yes	Yes	10,053,250	12,856,000	127.88	Yes
	33	Faridpur	Yes	Yes	11,750,000	8,560,597	72.86	Yes
	34	Gopalgonj	Yes	Yes	3,218,700	1.370,503	42.58	Yes
	35	Kushtia	Yes	Yes	9,525,000	6,366,005	66.83	Yes
	Total[b]				168,363,011	119,619,150	71.05	

PRAP = poverty reduction action plan, UGIIP = Urban Governance and Infrastructure Improvement (Sector) Project.
Notes:
[a] Mymensingh was not considered, as it was dropped from the project after it converted to a city corporation.
[b] Numbers may not sum precisely because of rounding.
Source: UGIIP Project Management Office.

Table A8.3: Status of Formation of Slum Improvement Committees to Implement Slum Improvement Activities of 35 Pourashavas, January–December 2020

UGIIP-3 Region	No.	Name of Pourashava[a]	SICs Formed in Selected Slums			Regular Meetings Held?	Activities Implemented by SIC Members with Effective Participation of all Members?	Number of Attendees		
			Number of SICs	Male	Female			Male	Female	Total
Bogura	1	Bera	6	6	78	Yes	Yes	67	692	759
	2	Chapainawabgonj	6	6	84	Yes	Yes	71	896	967
	3	Charghat	3	6	34	Yes	Yes	80	454	534
	4	Ishwardi	5	5	70	Yes	Yes	46	602	648
	5	Joypurhat	7	11	84	Yes	Yes	133	627	760
	6	Lalmonirhat	9	8	83	Yes	Yes	79	1,127	1,206
	7	Noagaon	6	0	77	Yes	Yes	32	807	839
	8	Nilphamari	5	5	53	Yes	Yes	50	525	575
	9	Panchagarh	6	6	84	Yes	Yes	45	722	767
	10	Shahjadpur	5	9	66	Yes	Yes	148	731	879
Cumilla	11	Bandarban	7	12	93	Yes	Yes	152	802	954
	12	Khagrachari	7	31	64	Yes	Yes	228	606	834
	13	Laksam	5	26	45	Yes	Yes	130	308	438
	14	Laxshmipur	7	26	67	Yes	Yes	148	577	725
	15	Nabinagar	5	24	48	Yes	Yes	197	599	796
	16	Rangamati	7	29	76	Yes	Yes	181	581	762
Magura	17	Benapole	4	3	85	Yes	Yes	22	478	500
	18	Chuadanga	4	5	58	Yes	Yes	40	437	477
	19	Jashore	9	8	24	Yes	Yes	36	939	975
	20	Kotalipara	5	5	63	Yes	Yes	32	388	420
	21	Magura	7	4	60	Yes	Yes	80	775	855

continued on next page.

Table A8.3 continued.

UGIIP-3 Region	No.	Name of Pourashava[a]	SICs Formed in Selected Slums			Regular Meetings Held?	Activities Implemented by SIC Members with Effective Participation of all Members?	Number of Attendees		
			Number of SICs	Members				Male	Female	Total
				Male	Female					
	22	Meherpur	6	11	64	Yes	Yes	81	710	791
	23	Rajbari	7	9	69	Yes	Yes	88	678	766
	24	Tungipara	1	0	30	Yes	Yes	7	69	76
Mymensingh	25	Chatak	2	15	45	Yes	Yes	158	148	306
	26	Habigonj	5	0	72	Yes	Yes	22	347	369
	27	Kishoreganj	7	7	90	Yes	Yes	63	513	576
	28	Moulvibazar	4	16	66	Yes	Yes	40	230	270
	29	Muktagacha	2	5	47	Yes	Yes	110	895	1,005
	30	Netrokona	6	26	58	Yes	Yes	78	190	268
	31	Sherpur	6	6	84	Yes	Yes	56	835	891
Additional Financing	32	Cox's Bazar	10	7	143	Yes	Yes	81	935	1,016
	33	Faridpur	10	36	114	Yes	Yes	188	899	1,087
	34	Gopalgonj	10	16	134	Yes	Yes	119	946	1,065
	35	Kushtia	10	4	146	Yes	Yes	120	1,680	1,800
Total			211	393	2,558			3,208	22,748	25,956

SIC = slum improvement committee, UGIIP = Urban Governance and Infrastructure Improvement (Sector) Project.

Note: [a] Mymensingh was not considered, as it was dropped from the project after it converted to a city corporation.

Sources: UGIIP Project Manangement. Unite.

More livable cities. Inhabitants of Hotat Para slum in Nilphamari *Pourashava* have a reason to smile. They enjoy new roads, toilets, street lights, and more community facilities built by and for the community from funds the *pourashava* earned for governance reforms achieved under the ADB-supported UGIIP-3 project.

MAYORS ON UGIIP

Benapole Mayor **Ashraful Alam Liton**

"I had no idea what to do once I was elected. I had no real way of connecting with the people on a citywide scale. There was a real disconnect... We were starting from scratch. We had absolutely nothing... UGIIP has helped in avoiding confrontations with the public because we work so closely with the public, the people. The UGIIP model makes politicians work for the public."

Chandpur Mayor **Md. Jillur Rahman**

"To be independent, a *pourashava* needs to be financially stable. Still, two-thirds of my municipality needs roads, drainage, and water facilities. Under UGIIP-2, we built a public market, which we later expanded, and a great revenue collection is coming from the market. There are 28 shops, and the total collection was Tk10 million (about $117,000). Then we added a second storey to the market and expanded it. We are getting rent every month, which we are using for municipal activities."

Ghorashal Mayor **Md. Shariful Haque**

"UGIIP-2 was like a light in our *pourashava*. We started dreaming for our *pourashava* and worked hard. We established three primary schools using the UGIIP-2 funding and constructed roads and drainage facilities. We always monitored and took serious action regarding violence against women. The child marriage rate has decreased significantly in our area because of the campaign we started during UGIIP-2."

Khagrachari Mayor **Nirmalendu Chowdhury**

"In 2012, when UGIIP first included Khargrachari *Pourashava*, our income was very limited. We had no other income. After giving staff salary, it was difficult to do any other development work. Because of the project, we have trained 975 women in sewing and 1,178 men and women in computer training. Our housing project under UGIIP is one of our most effective projects, and it has helped people in need. In this remote and hard-to-reach area, UGIIP development has brought drastic changes to residents' lives and is enabling us to earn greater from tax collection which, in return, is helping us to continue building the city."

Moulavibazar Mayor **Md. Fazlur Rahman**

"UGIIP-3 started in 2016, when I was first elected mayor, and I have been working on it since then. At every step, we were guided by the project, which is a crucial learning point for a municipality like us. We received required funding and contractors were able to give us quality work, and I made sure the work finished on time. We believe this was huge success. I was re-elected as mayor without much campaigning because our work with UGIIP spoke for itself…Now our tax collection rate is 95%, which is huge in a municipality like ours. I give 100% of the credit to UGIIP."

Narayanganj Mayor **Selina Hayat Ivy**

"UGIIP was our first experience working with an international project. I have learned almost everything through this project…Now we can manage big environmental projects and other investments because of our involvement with the ADB-financed UGIIP."

Nilphamari Mayor **Dewan Kamal Ahmed**

"Residents were concerned about infrastructure and development, but they had no idea how a *pourashava* is dependent on tax collection. But later, we started working with grassroots-level people, which has helped us to connect with our residents, and we have built relationships with them…We have built this city from almost nothing, and now I dream to make it a model *pourashava*."

Sherpur Mayor **Golam Mohammod Kibria**

"Until UGIIP started, we had no significant work to show. But UGIIP has helped us and taught us how to be transparent about income and expenses. This has helped us to fairly run this municipality. With the support of the government, we can barely develop infrastructural improvements…Last fiscal year, the tax collection rate was 87%. We had only a minimum tax collection before UGIIP."

Chapaynawabganj Mayor **Md. Mokhlasur Rahman**

"In 2014–2015, the *pourashava* collected a holding tax of Tk13.2 milion ($121,000). In 2020–2021, the *pourashava* collected more than Tk30 million (about $353,000), which is a great achievement, I believe. Our tax collection rate increased by 177%, and our non-holding tax collection has also increased by 82% in recent years."

Bhairab Mayor **Md. Iftakhar Hossain**

"Women have gained space in this *pourashava*. For example, there are now four female commissioners, and in our primary school, 80% are female teachers. We are focusing on education and mainly on girls' education."